MATHEMATICS FOR

GROB BASIC ELECTRONICS

This book is printed on recycled paper containing 10% post consumer waste.

BERNARD GROB

FOURTH EDITION

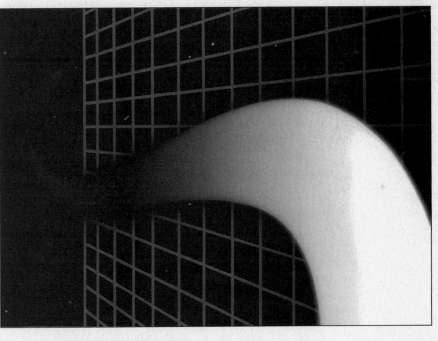

GLENCOE
McGraw-Hill

New York, New York   Columbus, Ohio   Mission Hills, California   Peoria, Illinois

This textbook has been prepared with the assistance of Publishing Advisory Service.

Mathematics for Grob Basic Electronics, Fourth Edition

Imprint 1995

Send all inquiries to:
Glencoe/McGraw-Hill
936 Eastwind Drive
Westerville, OH 43081

ISBN 0-02-800768-9

Printed in the United States of America.

5 6 7 8 9 10 11 12 13 14 15   MAL   99 98 97 96 95

CONTENTS

PREFACE

Mathematics for Basic Electronics provides students with the tools needed to solve problems in electricity and electronics. In the text and the examples, the terminology of electricity and electronics is used freely so the language becomes familiar. A chapter on computer mathematics covers the mathematical operations and the logic and symbols used in that field. However, the student is not expected to apply those technical principles to the solution of the problems.

Most students come to this book with a basic knowledge of arithmetic and the fundamental computational skills. They soon learn that the correct arithmetic manipulations of numbers does not always guarantee a correct answer. What these students need is training in how to apply the principles properly.

To deal with mathematical problems in electricity, electronics, and computers, a student must be able to:

- Keep track of the decimal point when multiplying and dividing decimal numbers.
- Work with fractions quickly and accurately.
- Manipulate reciprocals.
- Find powers and roots of a number (primarily squares and square roots).

These are the topics the student will concentrate on in the first four chapters. Chapter 5, on powers of 10, and Chap. 6, on logarithms, introduce essential computational aids in solving problems in electricity and electronics. Chapter 7 covers the metric system, focusing on metric prefixes and conversions between metric and U.S. Customary System (USCS) units.

Formulas and equations used to solve technical and scientific problems are usually stated in terms of symbolic letters, called literal numbers, which represent unknown quantities. The manipulation of expressions and formulas using these literal numbers is covered in Chap. 8. Chapters 9 and 10 cover methods of solving equations and simultaneous linear equations.

The analysis of many electrical circuits requires that the student be familiar with the measurements and properties of angles and triangles. Chapter 11, on trigonometry, explains the relationships between the sides and angles of a triangle and discusses the fundamental trigonometric functions of sine, cosine, and tangents.

Finally, in recognition of the early introduction of computers and especially digital systems in the study of electronics, Chap. 12 covers operations with binary (base 2) numbers and hexadecimal (base 16) numbers. Methods of converting numbers from one base to another are stressed. Also, the American Standard Code for Information Interchange (ASCII) is discussed as the bridge between binary notation and conventional alphanumeric symbols (letters, numbers, punctuation, etc.) used in communicating. Binary logic, logic gates, truth tables, boolean algebra, and the simplification of boolean expressions round out this introduction to computer mathematics.

Topics are divided into discrete sections, each stressing a single concept or operation. The individual steps have examples that illustrate how the concept or operation is used in an actual problem, followed by sets of practice problems. Answers to the practice problems are given at the end of each chapter. For the student, the dual advantages of this presentation are the immediate reinforcement of success with the correct answers and a minimum confusion of ideas. Each chapter concludes with a set of review problems, as a test of the principles covered in that chapter. Answers for odd-numbered review problems are at the end of the book.

There is no doubt that the electronic calculator can make computational work easier, but only if the student appreciates the principles underlying the calculations. Otherwise, the result can be an answer accurate to many decimal places but of the wrong value. Methods of using a calculator in the mathematical problems are explained in detail. For problems involving powers and roots, logarithms, and trigonometric functions, the scientific calculator is a necessity. To help explain calculator operations, the text uses special key symbols, such as $\boxed{x^2}$. Because calculators differ so greatly, symbols are used as guides only. Each student's calculator will have its individual characteristics.

This book is most effective with the author's text *Basic Electronics*. However, its coverage is independent of any electricity and electronics text. This book has one aim: to provide the mathematical principles needed to solve numerical problems in electricity and electronics. It is hoped that this aim is met.

Bernard Grob

1 DECIMAL NUMBERS AND ARITHMETIC

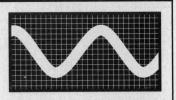

The system of numbers we generally use for counting and calculations is based on 10 digits. Including zero, the 10 digits are 0, 1, 2, 3, 4, 5, 6, 7, 8, and 9. Such a number system with 10 digits uses the *base 10*.

These values are decimal numbers, meaning that they are on a scale of tens, as the values increase by tens, hundreds, thousands, etc. Methods of using decimal numbers are explained in the following sections:

1-1 The Order of Places

In base 10 with 10 digits, the value of a digit can only go up to 9 because the first value is 0. In order to count values larger than 9, each position in a decimal number increases in value by a multiple of 10. The use of four places is shown in Fig. 1-1 on page 2. The places are:

Place a = units place. Values are 0, 1, 2, 3, 4, 5, 6, 7, 8, or 9.

Place b = tens place. Values are 0, 10, 20, 30, 40, 50, 60, 70, 80, or 90.

Place c = hundreds place. Values are 0, 100, 200, 300, 400, 500, 600, 700, 800, or 900.

Place d = thousands place. Values are 0, 1000, 2000, 3000, 4000, 5000, 6000, 7000, 8000, or 9000.

More places can be used, increasing to the left in multiples of 10, for ten thousands, hundred thousands, millions, etc.

We can use the number 3333 to illustrate the four places a, b, c, and d. The count is

3 units in place a for 3
3 tens in place b for 30
3 hundreds in place c for 300
3 thousands in place d for 3000

The total count, then, is 3000 and 300 and 30 and 3, for 3333 units.

Example Indicate the value in each place for the number 5432.

Answer The 2 in the right-hand place a is for 2 units.
The 3 for the next digit to the left is for tens in place b, which equals 30.
The 4 for the next digit to the left is for hundreds in place c, which equals 400.
The 5 for the next digit to the left is for thousands in place d, which equals 5000.
In summary, for this example of a decimal number,

$$5432 = 5000 + 400 + 30 + 2$$

Note that four places are used in this example. Also, a number in the thousands takes three places, while a number in the hundreds takes two places.

Practice Problems 1-A
Answers at End of Chapter

How many places are used in the following numbers?

1.	32	**5.**	16,666	**9.**	5
2.	302	**6.**	28,432,657	**10.**	55
3.	9999	**7.**	459	**11.**	555
4.	20	**8.**	10,000	**12.**	5555

1

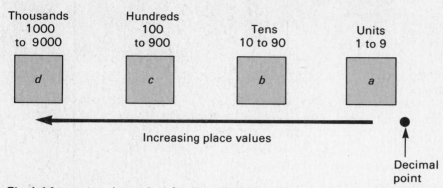

Fig. 1-1 Increasing place values for decimal numbers greater than 1. Places are indicated here as *a, b, c,* and *d* for four places increasing in multiples of 10.

Note that a count up to nine can be used in each place, as in 9999. Also, a zero is used to fill in a place to show that there is no count there, as in 302 and 20. In the number 302, there is no count for the tens. For 20, there is no count in the units place.

In writing large numbers (10,000 or more), it is standard practice to mark the thousands place and millions place with a comma; this is not done for numbers with four places. As examples, we have 50 or 500 or 5000 or 50,000 or 500,000 or 5,000,000. In some technical books, the comma is omitted and space is left between the thousands places. For example, 50 000 or 500 000 or 500 000 000.

Example Write the following quantities as a number and indicate the decimal point:

1 unit, 2 tens, 3 hundreds, and 4 thousands

Answer
1 unit	=	1
2 tens	=	20
3 hundreds	=	300
4 thousands	=	4000
	Number =	4321.

Practice Problems 1-B
Answers at End of Chapter

Give the value of the tens place in each number listed for Practice Problems 1-A.

1-2 The Decimal Point

Although it may not always be shown, the decimal point is just to the right of the units digit, as in Fig. 1-1. The number 5432 is actually

5432.

The value has not been changed in any way by indicating the decimal point. The decimal point simply marks where the places begin.

Places to the left of the decimal point are for numbers larger than 1. More places to the left mean a larger number. For instance, 2,000,000 is more than 2000.

An electronic calculator automatically shows the decimal point in the correct place. For numbers greater

Practice Problems 1-C
Answers at End of Chapter

Write each of the following quantities as a number and show the decimal point.

1. 2 tens plus 5 units
2. 2 hundreds plus 2 tens and 5 units
3. 5 hundreds plus 5 tens plus 5 units
4. 5 hundreds plus 5 units
5. 2 thousands plus 3 units
6. 8 thousands, 8 hundreds, 8 tens, and 8 units
7. 6 hundreds plus 6 units
8. 7 thousands plus 7 tens
9. 4 units
10. 4 tens
11. 4 hundreds
12. 4 thousands

---

than 1, the decimal point is after the last digit you punch in, which is the units place. For decimal fractions, you must punch in the decimal point so that it appears before the first digit punched in, which is the tenths place. As a comparison, if 2 is punched in first the number 2.0 appears, but if the decimal point is punched in first then 0.2 is displayed.

1-3 Decimal Fractions

Places to the right of the decimal point are used for fractional values that are less than 1 unit. As shown in Fig. 1-2, the place values decrease in multiples of tenths. One-tenth is 0.1, or ¹⁄₁₀. The number 0.1 is a *decimal fraction*, while ¹⁄₁₀ is a *common fraction*. Decimal fractions are generally used in scientific work because they are easier to use in calculations than common fractions.

For numbers less than 1, such as 0.1, the general practice is to write a zero to the left of the decimal point, to emphasize that there is no count in the units place.

Smaller decimal fractions use more places to the right of the decimal point. For the three places indicated in Fig. 1-2,

Place z = tenths place. Values are 0, 0.1, 0.2, 0.3, 0.4, 0.5, 0.6, 0.7, 0.8, or 0.9.
Place y = hundredths place. Values are 0, 0.01, 0.02, 0.03, 0.04, 0.05, 0.06, 0.07, 0.08, or 0.09.
Place x = thousands place. Values are 0, 0.001, 0.002, 0.003, 0.004, 0.005, 0.006, 0.007, 0.008, or 0.009.

More places to the right can be used for smaller decimal values. As an example, 0.000 005 is five millionths. Furthermore, more places to the right mean a smaller decimal fraction. For instance, 0.05, or ⁵⁄₁₀₀ is less than 0.5, or ⁵⁄₁₀. Note the zero in the units place before the decimal point. This is how decimal fractions are displayed on electronic calculators.

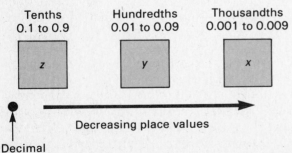

Fig. 1-2 Decreasing place values for decimal fractions less than 1. Places are indicated here as *z, y,* and *x* for three places decreasing in multiples of 10.

Example	Indicate the value in each place for the number 0.234.
Answer	The digit 2 in place *z* to the right of the decimal point means 2 tenths, or 0.2. The digit 3 in the next place *y* means 3 hundredths, or 0.03. The digit 4 in the next place *x* means 4 thousandths, or 0.004.

The complete decimal fraction, then, is

$$0.2 + 0.03 + 0.004 = 0.234$$

Practice Problems 1-D
Answers at End of Chapter

Write each value as a decimal fraction.

1. 6 tenths
2. 6 tenths plus 2 hundredths
3. 7 thousandths
4. 7 units plus 3 tenths
5. 2 tenths plus 5 thousandths
6. 9 tenths, 9 hundredths, and 9 thousandths
7. 7 millionths
8. 2 tenths and 7 millionths
9. 6 hundredths
10. 6 thousandths
11. 2 hundredths
12. 42 hundredths

A count up to nine can be used in each place, as in 0.999. But 9 is the highest count in any place.

Also, a zero is used to fill in a place to show that it has no count. For instance, in 0.205, there is no count for the hundredths place.

Practice Problems 1-E
Answers at End of Chapter

Give the value in the tenths place for the following decimal fractions.

1.	0.388	4.	0.088	7.	9.188	10.	0.4321
2.	0.288	5.	7.388	8.	9.088	11.	0.50
3.	0.188	6.	7.288	9.	0.1234	12.	0.55

1-4 Addition

The first step in adding decimal numbers is to line up the decimal points in a vertical column. Otherwise, you could be adding digits with different place values, which would result in the wrong count.

Example Add 321 and 45.

Answer The assumed decimal point is at the right of the units place. Then

$$
\begin{array}{r}
321 \\
+\ 45 \\
\hline
366 = \text{Sum}
\end{array}
$$

For two numbers, the given number is called the *augend*, the number to be added is the *addend*, and the result of the addition is the *sum*.

Example Line up the decimal points and add the three numbers 8432, 321, and 45.

Answer
$$
\begin{array}{r}
8432. \\
321. \\
45. \\
\hline
8798.
\end{array}
$$

This sum of 8798 can also be obtained by adding 8432 + 366 in the example above. Addition can be done in any order. For instance, 2 + 3 = 5 and 3 + 2 = 5 provide the same sum. The fact that addition can be done in any order is called the *law of commutation*.

Practice Problems 1-F
Answers at End of Chapter

1.	20 + 1 =	**5.**	762 + 215 =
2.	300 + 20 + 1 =	**6.**	762 + 21 =
3.	321 + 444 =	**7.**	8431 + 136 =
4.	1222 + 333 =	**8.**	421 + 112 + 23 =

The procedure used to add numbers greater than 1 also applies to adding decimal fractions. You still line up the decimal points.

Example Add 0.321 and 0.450.

Answer The decimal point is to the right of the zero in the units place, or to the left of the tenths place. Then

$$
\begin{array}{r}
0.321 \\
+\ 0.450 \\
\hline
+\ 0.771
\end{array}
$$

Notice that this sum has different digits from those in the sum for 321 + 45 = 366. The reason is that the place values go from left to right from the decimal point in a fraction, but they go from right to left in numbers greater than 1.

Practice Problems 1-G
Answers at End of Chapter

1. 0.4 + 0.03 =
2. 0.04 + 0.03 + 0.002 =
3. 0.24 + 0.35 =
4. 0.0002 + 0.0002 =
5. 0.0002 + 0.002 =
6. 0.52 + 0.25 =
7. 0.364 + 0.523 =
8. 0.724 + 0.241 + 0.03 =

So far, these examples of addition have not had a sum greater than 9 in any column. However, when the sum for a column is 10 or more, the result is an *overflow*, or *carry*, to the next column. With decimal numbers, a count of 10 or more cannot be shown in any column, but the excess is carried over to the next column to show the count. The column with a count of 10 or more has a *carry-out*. The next column to the left has a *carry-in*.

Example Add 838 and 725.

Answer
$$
\begin{array}{r}
838 \\
+\ 725 \\
\hline
1563
\end{array}
$$

In the units column at the right, note that 8 + 5 is actually 13. This is 10 + 3. Then 3 is in the units column, with a carry of 1 to the tens column. The sum in this column is 2 + 3 + carry of 1, which equals 6 without a carry for the tens column. In the hundreds column, 8 + 7 equals 15, which is 5 with a carry of 1 to the thousands column. Therefore, the correct count for this addition is one thousand, five hundred sixty three. The columns must be added from the units place at the decimal point moving left in order to put the carry into a larger place value.

Practice Problems 1-H
Answers at End of Chapter

1. $48 + 75 =$
2. $22 + 77 =$
3. $23 + 77 =$
4. $98 + 41 =$

5. $2432 + 123 =$
6. $2432 + 123 + 600 =$
7. $2432 + 1000 + 7000 =$
8. $4628 + 7777 + 8888 =$

When two numbers are added, the sum in any column cannot have a carry of more than 1. However, with more numbers, the carry can have any value.

The same method of carry-out and carry-in applies to the addition of decimal fractions.

> **Example** Add 0.827 and 0.645.
>
> **Answer**   0.827
>   + 0.645
>   ‾‾‾‾‾‾‾
>   1.472

The sum of $7 + 5$ in the thousandths column is actually 12 thousandths. This value is actually 2 thousandths and 1 hundredth, which is the carry of 1 to the hundredths column. In this column, the sum is 7 hundredths without any carry-out. Finally, $8 + 6$ in the tenths column is 14 tenths, which requires a carry of 1 past the decimal point to the units column. After all, 14 tenths is the same as 1 and 4 tenths, since 10 tenths equal 1 unit.

Note that the columns must be added from right to left in order to put the carry into a larger place value. This system is always used for the addition of columns, either with decimal fractions or numbers greater than 1.

Practice Problems 1-I
Answers at End of Chapter

The following problems combine decimal fractions and numbers greater than 1, with and without any carry for the addition.

1. $682 + 17 =$
2. $682 + 170 =$
3. $682 + 777 =$
4. $764 + 31.2 =$
5. $0.423 + 0.234 =$
6. $0.784 + 0.815 =$
7. $562 + 7.32 =$
8. $98 + 84 + 77 =$

9. $68.4 + 0.93 =$
10. $5,462,832 + 7 =$
11. $5,462,832 + 7,000,000 =$
12. $4625 + 7.843 =$

1-5 Subtraction

As in addition, before doing a problem in subtraction, make sure the decimal points in the two numbers are lined up.

> **Example** Subtract 23 from 98. The decimal point is assumed to be after the last digit in the units place.
>
> **Answer**   98
>   − 23
>   ‾‾‾‾
>   75   Difference

The given number, which is 98, is called the *minuend;* the number to be subtracted is the *subtrahend;* and the answer is the *difference.*

The same method applies to subtraction of decimal numbers.

> **Example** Subtract 0.341 from 0.875.
>
> **Answer**   0.875
>   − 0.341
>   ‾‾‾‾‾‾‾
>   0.534   Difference

Practice Problems 1-J
Answers at End of Chapter

1. $74 − 43 =$
2. $7.4 − 4.3 =$
3. $89.34 − 11.2 =$
4. $89.34 − 1.12 =$

5. $999 − 888 =$
6. $1575 − 1001 =$
7. $0.0054 − 0.0022 =$
8. $7,432,261 − 4,000,000 =$

So far, a smaller digit has always been subtracted from a larger digit in every column. When there is a larger digit in the subtrahend, however, it is necessary to *borrow* a 1 from the next higher place to the left. You borrow in the minuend, which is the number at the top.

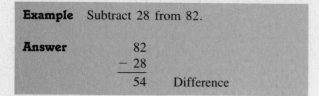

Example Subtract 28 from 82.

Answer
$$\begin{array}{r} 82 \\ -\ 28 \\ \hline 54 \end{array}\quad \text{Difference}$$

Note that in the units column, the 2 at the top was increased to 12 by borrowing 1 from the tens column. In this column, $12 - 8$ gives 4. The 8 in the tens column is decreased to 7, though, by the borrow. In the tens column, then, the subtraction is $7 - 2$, for 5. The resulting difference is $82 - 28 = 54$.

As a proof, the difference can be added to the subtrahend to get the original minuend. Then we have $54 + 28 = 82$.

The principle of the borrow for subtraction is the opposite of that of the carry in addition. However, the borrow can only be a 1, not more, for any of the place columns.

The borrow method for subtraction also applies to decimal fractions. Always subtract the columns starting with the smallest place value first and then moving to the left. This way, if a borrow is necessary, it comes from the next higher place value.

To summarize the different possibilities, these rules always apply in addition and subtraction:

a. Line up the decimal points.
b. Do the columns from right to left.

The rules are the same for decimal fractions and numbers greater than 1.

As a result, either a carry for addition or a borrow for subtraction will always be in the correct decimal place. Furthermore, this method is correct even when there is no carry or borrow.

Practice Problems 1-K
Answers at End of Chapter

1. $578 - 263 =$
2. $578 - 293 =$
3. $0.074 - 0.022 =$
4. $0.074 - 0.026 =$
5. $1625.02 - 0.19 =$
6. $1.10 - 1.02 =$
7. $1575 - 500 =$
8. $1575 - 800 =$
9. $48.33 - 21.88 =$
10. $62.1 - 31.9 =$
11. $422 - 188 =$
12. $6,000,000 - 288.6 =$

A special case arises when a larger number is subtracted from a smaller number. For instance, what is the difference when 54 is subtracted from 22, that is, $22 - 54$? The method here is to subtract the smaller number from the larger number and place a negative sign in front of the answer. Then $22 - 54 = -32$. The use of negative numbers is explained in Chap. 2.

1-6 Multiplication

In order to define some names, we can use the example of $3 \times 2 = 6$. The 3 is the *multiplicand,* 2 is the *multiplier,* and the answer of 6 is the *product.* Both the multiplicand and the multiplier are *factors* of the product.

With two factors, one factor can be divided into the product to yield the other factor.

Multiplication can be done in any order, as $2 \times 3 = 6$ is the same as $3 \times 2 = 6$.

When more than two numbers are to be multiplied, just continue the multiplication using the intermediate products. As an example, $2 \times 3 \times 4 = 24$. This is actually done in steps: $2 \times 3 = 6$; the 6×4 then is equal to 24.

Multiplication is basically a process of continued addition. For instance, 2×3 really means add 2 units 3 times. Then the result is $2 + 2 + 2 = 6$, the same as $2 \times 3 = 6$.

To set up a problem in multiplication with larger numbers, it is not necessary to line up the decimal points. For convenience, though, the right-hand digits of the two factors are aligned, as shown in the next example. The steps in multiplication are:

1. Multiply all digits in the multiplicand by the right-hand digit in the multiplier. The result is a *partial product.*
2. Multiply all digits in the multiplicand by the next digit to the left in the multiplier. Each partial product is lined up with the digit used in the multiplier.
3. Continue until all digits in the multiplier have been used to obtain partial products.
4. Add the partial products and locate the decimal point.

Example Multiply 13.4 by 2.1.

Answer This problem can be arranged in either of two ways:

$$\begin{array}{r} 13.4 \\ \times\ 2.1 \end{array}\quad \text{or} \quad \begin{array}{r} 2.1 \\ \times\ 13.4 \end{array}$$

Both methods will produce the same answer, but using the arrangement shown at the left requires fewer calculations because the multiplier has fewer digits.

To carry through the multiplication in this example, multiply each digit of the first number in order by the 1 in the second number (the multiplier). Line up the right-hand digit of the product with the 1. Ignore the decimal points at this stage.

$$
\begin{array}{r}
1\,3.4 \\
\times\ 2.1 \\
\hline
1\,3\,4
\end{array}
$$
Partial product

Next multiply each digit in the top number by the 2 in the multiplier. Again, ignore the decimal point. Line up the second product with the 2 in the multiplier, because now this is the digit you are using for multiplication.

$$
\begin{array}{r}
1\,3.4 \\
\times\ 2.1 \\
\hline
1\,3\,4 \\
2\,6\,8
\end{array}
$$
Partial products

Finally, add the two partial products:

$$
\begin{array}{r}
1\,3.4 \\
\times\ 2.1 \\
\hline
1\,3\,4 \\
2\,6\,8 \\
\hline
2\,8.1\,4
\end{array}
$$
Partial products

Answer

To locate the position of the decimal point in the answer, add the number of digits to the right of the decimal point in each of the two numbers that were multiplied. The first number has one digit to the right of the decimal point; the second number also has one digit to the right. Therefore, the final product will have two digits to the right of the decimal point. The answer is 28.14. The decimal points were lined up in this example because both factors have the same number of digits to the right of the decimal point.

However, it is not necessary to line up the decimal points in multiplication. For example, $134 \times 2.1 = 281.4$.

The multiplication of factors can be indicated in any of the following four ways:

$$6 \times 3 = \qquad 6 \cdot 3 = \qquad (6)(3) = \qquad 6(3) =$$

The dot for a product has the same meaning as the multiplication sign. Also, parentheses around numbers indicate that they are to be multiplied when there is no $+$ or $-$ sign to show addition or subtraction. If parentheses are not shown, though, digits are just part of a number. For instance, 63 is sixty-three but (6)(3) or 6(3) is 6×3 or 18.

It should be noted that in algebra, with digits and letters such as a, b, c, x, y, and z for literal numbers, the combination $6a$ for a specific value means $6 \times a$. See Chap. 8 for more details of algebra.

Practice Problems 1-L
Answers at End of Chapter

1. $3.2 \times 3 =$
2. $32 \times 0.3 =$
3. $42 \times 0.2 =$
4. $12.3 \times 2.2 =$
5. $400 \times 0.2 =$
6. $0.2 \times 400 =$
7. $11.1 \times 9 =$
8. $420 \times 2.2 =$
9. $13.4 \times 2.1 \times 2 =$
10. $(27.3)(2) =$
11. $(14)(12) =$
12. $4 \times 4 \times 4 =$
13. $4 \cdot 8 =$
14. $(4)(8) =$

1-7 Division

The form in which a division problem is given usually depends on the amount of numbers involved. If we need to divide 6 by 3, as a simple example, the problem could be written as

$$6/3 = 2 \qquad \text{or} \qquad \frac{6}{3} = 2 \qquad \text{or} \qquad 6 \div 3 = 2$$

We know the answer is 2 because we have memorized the fact that $2 \times 3 = 6$. In this example of $6 \div 3 = 2$, the 6 is the *dividend*, 3 is the *divisor*, and the answer of 2 is the *quotient*. Note the methods of indicating division: a slanted line (sometimes called a shilling or slash), a horizontal line, and a division sign, shown from left to right in the example above.

When division is shown with a horizontal line, the dividend is at the top and the divisor is at the bottom. In this case the division problem is in the form of a fraction. The dividend at the top is called the *numerator* of the fraction and the divisor at the bottom is the *denominator*. More details on fractions are explained in Chap. 3.

It is important to note that although it does not matter which of the two numbers is the multiplier in the operation of multiplication, the divisor and dividend cannot be interchanged. For instance, $6 \div 3 = 2$, but $3 \div 6$ is 3/6, 1/2, or 0.5.

When the divisor has two or more digits, the method of *long division* is usually necessary.

Example   Divide 180 by 12. This type of problem can be written as

$$12\overline{)180}$$

Answer   The quotient is obtained by first testing the dividend to find the lowest number into which the divisor can go. In this example, the divisor, 12, can go into 18 once, leaving a remainder of 6. This is combined with the next place in the dividend.

The new number, 60 in this example, is divided by 12, and the answer becomes the next digit in the quotient. The entire division is shown as follows:

$$
\begin{array}{r}
15 \\
12\overline{)180} \\
(1 \times 12 =) \quad 12 \\
\overline{60} \\
(5 \times 12 =) \quad 60 \\
\overline{0}
\end{array}
$$

Continue the process until an answer digit is above each of the remaining dividend digits. Any number left to be divided is called the *remainder*. The remainder is zero in this example. The answer is exactly 15 with a remainder of zero because $12 \times 15 = 180$ exactly.

When there is a decimal point in the dividend, the decimal point in the quotient is lined up with the decimal point in the dividend.

Example   Divide 38.4 by 12.

Answer   This problem is solved by long division as follows:

$$
\begin{array}{r}
3.2 \\
12\overline{)38.4} \\
3\,6 \\
\overline{2\,4} \\
2\,4 \\
\overline{0}
\end{array}
$$

There should not be a decimal fraction in the divisor. Otherwise, there may be a question of where to locate the decimal point in the quotient. Always move the decimal point to the right in the divisor when necessary to eliminate any decimal fraction, and move the decimal point the same number of places in the dividend. The quotient's value is not changed by this procedure.

Example   Divide 3.84 by 1.2.

Answer   Move the decimal point one place to the right in both the divisor and the dividend. Then

$$1.2\overline{)3.84} \quad \text{becomes} \quad 12\overline{)38.4}$$

The answer here is 3.2, the same as in the previous example.

You can move the decimal point because you are actually multiplying the divisor and dividend by the same number. As a result, the quotient is the same.

Practice Problems 1-M
Answers at End of Chapter

1.	$180 \div 12 =$	**6.**	$2472 \div 12 =$
2.	$168 \div 14 =$	**7.**	$247 \div 13 =$
3.	$165 \div 15 =$	**8.**	$340 \div 17 =$
4.	$238 \div 17 =$	**9.**	$196 \div 14 =$
5.	$238 \div 14 =$	**10.**	$210 \div 14 =$

Practice Problems 1-N
Answers at End of Chapter

1.	$16.8 \div 12 =$	**6.**	$168 \div 1.4 =$
2.	$16.8 \div 14 =$	**7.**	$16.8 \div 1.4 =$
3.	$16.5 \div 15 =$	**8.**	$247.2 \div 0.12 =$
4.	$24.72 \div 12 =$	**9.**	$165.6 \div 1.8 =$
5.	$168 \div 1.2 =$	**10.**	$165.6 \div 3.6 =$

In some cases of division, the answer will not be an exact whole number. Then zero can be added to the right of the decimal point in the dividend, to as many places as necessary.

Example Divide 228 by 16.

Answer This is solved as follows:

$$
\begin{array}{r}
14.25 \\
16\overline{)228.00} \\
\underline{16} \\
68 \\
\underline{64} \\
40 \\
\underline{32} \\
80 \\
\underline{80} \\
0
\end{array}
$$

When the dividend is not exactly divisible by the divisor, a remainder of zero cannot be obtained. Then you have these three possibilities:

1. When the remainder is less than one-half the divisor, consider the remainder as zero and drop it.
2. When the remainder is more than one-half the divisor, increase the last digit in the quotient by one.
3. When the remainder is exactly one-half the divisor, make the next and last digit in the quotient a 5. This corresponds to ½, or 0.5, for the last decimal place.

Practice Problems 1-O
Answers at End of Chapter

1.	$396 \div 12 =$	7.	$528 \div 15 =$
2.	$39.6 \div 12 =$	8.	$6327 \div 17 =$
3.	$3.96 \div 12 =$	9.	$242 \div 363 =$
4.	$39.6 \div 1.2 =$	10.	$121 \div 363 =$
5.	$175 \div 25 =$	11.	$4672 \div 17.8 =$
6.	$0.175 \div 0.25 =$	12.	$6.431 \div 0.785 =$

1-8 Moving the Decimal Point

In our number system based on ten, we can multiply by 10 or 100 or 1000 merely by moving the decimal point one, two, or three places to the right.

Example Multiply 4.87 by 1000.

Answer We can multiply this out by the usual method:

$$
\begin{array}{r}
1000 \\
\times\ 4.87 \\
\hline
7000 \\
8000 \\
4000 \\
\hline
4870.00
\end{array}
$$

Notice that the original digits 4, 8, and 7 were unchanged. The decimal point just moved three places to the right. Then we have

$$4.87 \times 1000 = 4870$$

The rule for multiplying any number by a power of 10 (that is, 10, 100, 1000, 100,000, etc.) is to move the decimal point to the right as many places as there are zeros in the multiplier. When multiplying by 10, move the decimal point one place to the right; when multiplying by 100, move the decimal point two places to the right; when multiplying by 1000, move the decimal point three places to the right; and so on.

Practice Problems 1-P
Answers at End of Chapter

1.	$7 \times 100 =$	6.	$4.82 \times 10 =$
2.	$7.82 \times 100 =$	7.	$9.64 \times 1000 =$
3.	$78.2 \times 100 =$	8.	$96.4 \times 1000 =$
4.	$782 \times 100 =$	9.	$8.5 \times 100 =$
5.	$48.2 \times 10 =$	10.	$85 \times 10 =$

This rule also applies to decimal fractions. The decimal point moves to the right for multiplication, whether the number is greater than 1 or less than 1.

Example Multiply 0.0042 by 100.

Answer Since 100 has two zeros, the decimal point is moved two places to the right.

$$0.0042 \times 100 = 0.42$$

Practice Problems 1-Q
Answers at End of Chapter

1.	$0.07 \times 10 =$	**6.**	$0.0234 \times 10 =$
2.	$0.07 \times 100 =$	**7.**	$0.896 \times 100 =$
3.	$0.007 \times 1000 =$	**8.**	$0.0004 \times 1000 =$
4.	$0.007 \times 10 =$	**9.**	$0.532 \times 100 =$
5.	$0.234 \times 10 =$	**10.**	$0.0532 \times 1000 =$

It is not necessary to retain more than one zero to the left of the decimal point. A single zero in the units column is enough to indicate that the number is less than 1. For instance, 0000.4 is the same as 0.4.

In the case of division by powers of 10, the decimal point is moved to the left as many places as there are zeros in the divisor. This procedure is the opposite of the method for multiplication.

Example   Divide 642 by 100.

Answer   Since 100 as the divisor has two zeros, the decimal point in the dividend is moved two places to the left.

$$642 \div 100 = 6.42$$

This method of division also applies to decimal fractions.

Example   Divide 0.42 by 100.

Answer   Move the decimal point two places to the left.

$$0.42 \div 100 = 0.0042$$

In summary, the rules for moving the decimal point are as follows:

For multiplication, move to the right ➡.
For division, move to the left ⬅.

Practice Problems 1-R
Answers at End of Chapter

1.	$200 \div 100 =$	**4.**	$0.57 \times 100 =$
2.	$580 \div 100 =$	**5.**	$0.57 \div 100 =$
3.	$5432 \div 1000 =$	**6.**	$0.794 \div 10 =$

7.	$0.036 \times 100 =$	**9.**	$379 \times 10 =$
8.	$0.036 \div 100 =$	**10.**	$379 \div 10 =$

1-9   The Percent (%) Sign

The percent (%) sign means hundredths, as 1% = 0.01. For example, 50% means 50 hundredths, 0.50, $^{50}/_{100}$, or one-half. To change a value in percent to the equivalent decimal value, substitute 0.01 for the percent sign and multiply. As examples:

$1\% = 1 \times 0.01 = 0.01$	$17\% = 17 \times 0.01 = 0.17$
$2\% = 2 \times 0.01 = 0.02$	$24\% = 24 \times 0.01 = 0.24$
$5\% = 5 \times 0.01 = 0.05$	$50\% = 50 \times 0.01 = 0.50$
$9\% = 9 \times 0.01 = 0.09$	$92\% = 92 \times 0.01 = 0.92$

Multiplying by 0.01 or $^{1}/_{100}$ is the same as dividing by 100, which means moving the decimal point two places to the left.

Going the other way, to change from a decimal fraction to a percent:

1. Have the number in a form that has at least two places to the right of the decimal point. For instance, state 0.4 as 0.40.
2. Multiply by 100%. This multiplier is actually 1, since $100 \times 0.01 = 1$.

Examples of changing from a decimal fraction to percent are:

$$0.01 = 0.01 \times 100\% = 1\%$$
$$0.02 = 0.02 \times 100\% = 2\%$$
$$0.05 = 0.05 \times 100\% = 5\%$$
$$0.17 = 0.17 \times 100\% = 17\%$$
$$0.24 = 0.24 \times 100\% = 24\%$$
$$0.5 = 0.50 \times 100\% = 50\%$$

Multiplying by 100 is the same as moving the decimal point two places to the right.

The following values are less than 1% because the decimal values are smaller than 0.01:

$$0.001 = 0.001 \times 100\% = 0.1\% \quad \text{or}$$
$$0.1\% = 0.1 \times (0.01) = 0.001$$

$$0.002 = 0.002 \times 100\% = 0.2\% \quad \text{or}$$
$$0.2\% = 0.2 \times (0.01) = 0.002$$

$$0.005 = 0.005 \times 100\% = 0.5\% \quad \text{or}$$
$$0.5\% = 0.5 \times (0.01) = 0.005$$

Practice Problems 1-S
Answers at End of Chapter

Change the following decimal values to percentages.

1.	0.02	**5.**	0.4	**9.**	0.34
2.	0.34	**6.**	0.08	**10.**	0.6
3.	0.52	**7.**	0.003	**11.**	0.5
4.	0.78	**8.**	0.99	**12.**	0.1

Practice Problems 1-T
Answers at End of Chapter

Change the following percentages to decimal values.

1.	2%	**5.**	40%	**9.**	34%
2.	34%	**6.**	8%	**10.**	60%
3.	52%	**7.**	0.3%	**11.**	50%
4.	78%	**8.**	99%	**12.**	10%

1-10 Squares and Square Roots

The square of a number is that number multiplied by itself. When we say "square 5" or "5 squared," it just means 5×5, for the answer of 25. This is generally written as $5^2 = 5 \times 5 = 25$.

The 5 here is the base number. The 2 written above is the exponent or power of the base number. For the exponent 2, the base number is squared or raised to the second power. Any number can be the base. As examples:

$$3^2 = 3 \times 3 = 9$$
$$10^2 = 10 \times 10 = 100$$

The square root of a number is that number, which when multiplied by itself equals the original number. A symbol called a radical sign, $\sqrt[2]{}$, shows that we want to find a root. The 2 indicates a square root. However, if no number is shown, it is understood to mean a square root.

Example $\sqrt{25} = 5$

Answer The proof is that $5 \times 5 = 25$. Therefore, 5 is the square root of 25.

The purpose of using exponents is to have a shortcut method for continued multiplication of the same number. Actually, any number can be the base, raised to any power. These operations are described for cubes as the third power, and for all powers and roots in general, in Chap. 4. Exponents are especially useful as powers of 10 to keep track of the decimal point for very large or very small numbers, as described in detail in Chap. 5.

Practice Problems 1-U
Answers at End of Chapter

Find the square or square root of the following values.

1.	9^2	**11.**	$\sqrt{81}$
2.	8^2	**12.**	$\sqrt{64}$
3.	7^2	**13.**	$\sqrt{49}$
4.	6^2	**14.**	$\sqrt{36}$
5.	5^2	**15.**	$\sqrt{25}$
6.	4^2	**16.**	$\sqrt{16}$
7.	3^2	**17.**	$\sqrt{9}$
8.	2^2	**18.**	$\sqrt{4}$
9.	1^2	**19.**	$\sqrt{1}$
10.	0^2	**20.**	$\sqrt{0}$

Considerable calculating time and labor can be saved if the squares and square roots shown above are memorized. For more difficult problems, longhand methods must be applied or, best of all, an electronic calculator used to do the work. Note that for zero, its square, or any power, or any root is still zero. Also, any power or root of 1 is still 1.

Almost all scientific or technical calculators contain a $\boxed{x^2}$ key for finding squares and a $\boxed{\sqrt{}}$ key for finding square roots. For the square root, first you must press the second function $\boxed{\text{2ndF}}$ key when there is not a separate $\boxed{\sqrt{}}$ key. These keys are pressed after the given number is entered.

Practice Problems 1-V
Answers at End of Chapter

Using an electronic calculator, find the square or square root of the following values.

1.	$(14)^2$	**5.**	$(19)^2$	**9.**	$\sqrt{2916}$
2.	$(23)^2$	**6.**	$(25)^2$	**10.**	$\sqrt{153,664}$
3.	$(54)^2$	**7.**	$\sqrt{196}$	**11.**	$\sqrt{625}$
4.	$(392)^2$	**8.**	$\sqrt{529}$	**12.**	$\sqrt{10,000}$

1-11 Average Value

When a quantity has different values at different times, it is useful to consider one value as typical in order to have a specific measure. A value used very often is the *arithmetic average*. This value equals the sum of all the values divided by the number of values.

Example The test scores are 70, 80, and 90. Find the student's average score.

Answer To find the average, add the test scores and divide by the number of scores:

$$
\begin{array}{r} 70 \\ 80 \\ 90 \\ \hline 240 \end{array}
\qquad
\begin{array}{r} 80 \\ \hline 3)\overline{240} \end{array}
$$

The average score is 80. In this particular case it happens to be the middle, or *median*, score also. The arithmetic average is also called the arithmetic *mean*.

The arithmetic average is not always exactly in the middle, though, as shown by the next example.

Example Find the average of 60, 70, 80, and 80.

Answer For the addition,

$$60 + 70 + 80 + 80 = 290$$

For the division,

$$
\begin{array}{r} 72.5 \\ \hline 4)\overline{290.0} \end{array}
$$

The average of 60, 70, 80, and 80 is therefore 72.5.

Average calculations can misrepresent the actual physical conditions that are present in a problem, although they may be mathematically correct. If one number is very much different from the others in a group, the average value may be unduly weighted toward this value.

Example The following voltage readings were made in a circuit. Find the average voltage.

110, 112, 115, 115, 27

Answer Following the usual procedure, we find the average:

$$
\begin{array}{r} 110 \\ 112 \\ 115 \\ 115 \\ 27 \\ \hline 479 \end{array}
$$

$$
\begin{array}{r} 95.8 \\ \hline 5)\overline{479.0} \end{array}
$$

Although our calculations found an average value of 95.8 volts (V), this is not really the condition of the circuit. It is obvious that the 27-V reading has brought the average down below a reasonable value. Therefore, in finding an arithmetic average that must relate to some real physical condition, such as the readings in an experiment, it may be necessary to discard any value that is completely different from the others in the group. It is a good idea to check on what caused the unreasonable value.

Practice Problems 1-W
Answers at End of Chapter

Find the arithmetic average for the following groups of values.

1. 13, 8, and 9
2. 70, 80, 90, and 95
3. 0.5, 0.7, 0.4, and 0.8
4. 1.3, 0.8, and 0.9
5. 2, 4, 5, and 6
6. 3, 5, 7, and 8
7. 0 V, 33 V, 50 V, 78 V, 88 V, 97 V, and 100 V (*Note:* V is for volts.)
8. 100 Ω, 75 Ω, 83 Ω, 1000 Ω (*Note:* The Ω is for ohms of resistance.)
9. 4, 6, 8, 10, 11, and 12
10. 3, 4, 6, 8, 10, and 12

1-12 Root-Mean-Square (RMS) Value

In some applications, the squares of the individual values are important for the average. An example is electric power, which is often calculated using the formula I^2R. This means the power depends on the square of the current intensity I. Alternating current and voltage are usually specified by their *root-mean-square,* or *rms,* value.

An rms value is derived by taking the square of each individual value, adding the squares, and dividing the sum by the number of values. This results in a mean of the squares. The square root of this mean value is the rms value. In summary, we can think of rms as an abbreviation for the square root of the mean value of the squares of the individual values.

Example Find the rms value for 2, 3, and 4.

Answer The squares are 4, 9, and 16. Their sum is 29. The mean or average value of the sum of the squares is

$$\frac{29}{3} = 9.67$$

The square root of 9.67 is equal to 3.11. This answer of 3.11, then, is the rms value for 2, 3, and 4.

Practice Problems 1-X
Answers at End of Chapter

Find the rms value for the following groups of values.

1. 3, 4, 5, 6, and 7
2. 5, 7, 9, and 10
3. 2, 3, and 3
4. 20, 30, and 40
5. 4, 5, and 6
6. 0.4, 0.5, and 0.6
7. 2, 5, 7, and 10
8. 0, 1, 2, 3, 4, 5, 6, 7, 8, and 9
9. 1, 2, 3, 4, and 5
10. 2, 4, and 6
11. 3, 6, and 9

1-13 Order of Operations and Signs for Grouping Numbers

In a problem that combines different operations, they are done in the following order:

1. Powers and roots are calculated first.
2. Multiplication and division are done next.
3. Addition and subtraction are done last.

Do the calculations in this sequence, from left to right for each step in the calculation. Note that the highest order of difficulty, powers and roots, is first and the lowest order, addition and subtraction, is last. Some examples are:

$$3^2 \times 4 = 9 \times 4 = 36$$
$$6 \times 2 + 3 = 12 + 3 = 15$$
$$3^2 \times 4 + 5 = 9 \times 4 + 5 = 36 + 5 = 41$$

Practice Problems 1-Y
Answers at End of Chapter

1. $6 \times 2 + 4 =$
2. $6 \div 2 + 4 =$
3. $4^2 + 5 =$
4. $3^2 \times 6 + 5 =$
5. $6^2 \div 2 + 8 =$
6. $\sqrt{16} \times 4 - 3 =$
7. $\sqrt{9} \div 3 + 1 =$
8. $9 \times 2 + 4 - 2 =$

In order to clarify the sequence that should be used in combining numbers, especially with literal values in algebra, the following mathematical symbols or signs are often used for grouping:

Parentheses ( ) Do this first.
Brackets [ ] Do this second.
Braces { } Do this third.

Example $(2 + 3)^2 + 4$
$= 5^2 + 4 = 25 + 4 = 29$

Example $[3(2 + 3)^2 + 4]$
$= [3(5)^2 + 4] = [3(25) + 4]$
$= [75 + 4] = 79$

When a group has multiplications and divisions or powers and roots, these operations must be completed before the groups can be added or subtracted. As examples:

$$4 + (4 \times 2) = 4 + 8 = 12$$
$$35 - (2 + 3)^2 = 35 - (5)^2 = 35 - 25 = 10$$

For division, the fraction bar is a sign of grouping. You cannot divide until all the additions or subtractions are combined in either the numerator or denominator or both.

Example $2 + \dfrac{4+5}{3} = 2 + \dfrac{9}{3} = 2 + 3 = 5$

Practice Problems 1-Z
Answers at End of Chapter

1. $6 + (2 \times 3) =$
2. $(14 \times 2) - 20 =$
3. $\dfrac{8+12}{5} + 3 =$
4. $[2 + \frac{1}{2}(3 + 1) - 4] =$
5. $8(3 + 6) =$
6. $4 + (8 - 3)^2 =$
7. $3(2 + 2 - 1) =$
8. $\left(\dfrac{4+8}{6} + 3\right) \times 6 =$
9. $6 + (2 \times 4) =$
10. $(3 \times 8) - 9 =$

1-14 Rounding Off a Number

Sometimes we want to "round off" numbers because we do not need the accuracy indicated by all the digits. This is especially true with electronic calculators. There is no need for a value of 36.241 102 31 V, as an example, when the closest we can measure it with an analog voltmeter is probably 36.24 V.

Example Round off 469 to the nearest tens place.

Answer The answer here is 470, but let us see why. Just dropping the 9 in the units place gives the number 460, which is 9 less than the actual value. On the other hand, the 9 in the units place is very close to 10 for the next digit in the tens place. Therefore, raising the 6 to 7 in the tens place for the value of 470 makes the number only 1 more than the original. Thus 469 accurate to the nearest tens place is 470.

Example Round off 461 to the nearest tens place.

Answer The answer here is 460. The 1 in the units place can be changed to zero, without raising the digit in the tens place, as 460 is closer to the original number than 470.

The general rules are as follows:

1. When the digit dropped is 6 or more, raise the last digit by 1.
2. When the digit dropped is 4 or less, keep the last digit the same.

It is important to remember that the digit dropped must be replaced by a zero. This procedure keeps the same number of decimal places in the number.

Examples Rounded off to the nearest hundreds place:

$$4611 = 4600$$
$$4644 = 4600$$
$$4664 = 4700$$
$$4691 = 4700$$

All four examples are rounded off to two significant figures, which is the number of digits other than zero before the decimal place in the rounded number.

When the digit to be dropped is 5, there are two possibilities:

1. The 5 is followed by digits more than zero. This tips the value to more than one-half the decimal place. Then raise the digit before the 5 by 1.
2. The 5 is followed by one or more zeros. This value is exactly one-half the decimal place. Then you can raise the previous digit only if it becomes an even number.

Examples Rounded off to the nearest hundreds place:

$$4651 = 4700 \qquad 4750 = 4800$$
$$4659 = 4700 \qquad 4650 = 4600$$

When rounding off decimal fractions less than 1, the procedure is the same. However, the zeros following the rounded digits can be eliminated, because they do not affect the value.

Examples Rounded off to the nearest hundredths place:

$$0.4611 = 0.4600 = 0.46$$
$$0.4664 = 0.4700 = 0.47$$
$$0.4651 = 0.4700 = 0.47$$

Practice Problems 1-AA
Answers at End of Chapter

Round off the following to the number of significant digits shown in brackets.

1.	1279 [3]	**5.**	482 [1]
2.	1271 [2]	**6.**	486 [2]
3.	1275 [3]	**7.**	485 [2]
4.	0.1277 [3]	**8.**	0.3333 [3]

1-15 When a Zero Can Be Dropped

We can drop any zero that does not change the numerical value. As an example, 0.72 is the same as 0.720. Note that we keep the zero in the units place of a decimal fraction, just for show. It does not change the numerical value, but this zero emphasizes where the decimal point is. For a different example, you cannot drop the zero in 720 because 72 is not the same as 720.

The question usually arises with decimal fractions. You can usually drop any zeros after the last digit to the right in decimal fractions except where the zero is used to indicate the accuracy of the number.

Examples

$$0.432000 = 0.432$$
$$0.7200 \;\;\;= 0.72$$
$$0.40 \;\;\;\;\;= 0.4$$

All the zeros dropped here are after the last digit in the decimal fraction. This rule would not apply to numbers

greater than 1. For instance, in 432,000, 7200, or 40, no zeros could be dropped.

In the decimal fraction 0.707 you cannot drop the middle zero because it is before the last digit. Dropping this zero would change the numerical value to 0.77, which is more than 0.707.

Practice Problems 1-BB
Answers at End of Chapter

Answer yes or no if any zero can be dropped in the following numbers. (For decimal fractions, do not drop the zero in the units place.)

1.	900	**5.**	4020
2.	0.900	**6.**	4.020
3.	0.6370	**7.**	0.400001
4.	0.707	**8.**	0.410000

1-16 Evaluation of Formulas

In order to evaluate the unknown factor in a formula, just substitute the known values for the letters in the formula. As an example, we calculate the voltage from the Ohm's law formula $V = IR$. The IR means $I \times R$. The current I is in amperes, the resistance R is in ohms, and the voltage V is in volts. Let $I = 2$ amperes (A) and $R = 4$ ohms (Ω). Then

$$V = IR$$
$$= 2\,\text{A} \times 4\,\Omega$$
$$V = 8\,\text{V}$$

Practice Problems 1-CC
Answers at End of Chapter

Using the Ohm's law formula $V = IR$, calculate the voltage for the following values of I in amperes and R in ohms.

1.	$I = 2, R = 5$	**6.**	$I = 0.000\,002,$
2.	$I = 5, R = 2$		$R = 10,000$
3.	$I = 0.003,$	**7.**	$I = 0.003, R = 4,000,000$
	$R = 3000$	**8.**	$I = \sqrt{2}, R = 10$
4.	$I = 0.003,$	**9.**	$I = 0.05, R = 1000$
	$R = 100$	**10.**	$I = 0.07, R = 1000$
5.	$I = 10, R = 47$		

The Ohm's law formula can also be used as $I = V/R$ to find the current I when V and R are the known values.

As an example, with 8 V across 4 Ω,

$$I = \frac{V}{R} = \frac{8 \text{ V}}{4 \text{ } \Omega} = 2 \text{ A}$$

Practice Problems 1-DD
Answers at End of Chapter

Using the Ohm's law formula $I = V/R$, calculate the current I in amperes for the following values of V in volts and R in ohms.

1. $V = 10$, $R = 5$
2. $V = 10$, $R = 2$
3. $V = 9$, $R = 3000$
4. $V = 0.3$, $R = 100$
5. $V = 470$, $R = 47$
6. $V = 0.02$, $R = 10,000$
7. $V = 12,000$, $R = 4,000,000$
8. $V = 14.14$, $R = 10$
9. $V = 12$, $R = 1000$
10. $V = 12$, $R = 2000$

A similar application is using the power formula $P = I^2R$, where P is the power in watts (W), for I in amperes and R in ohms. The I^2R means $I^2 \times R$. As an example, we can calculate the power for the values of $I = 3$ A and $R = 4$ Ω. Then

$$
\begin{aligned}
P &= I^2R \\
&= (3)^2 \times 4 \\
&= 9 \times 4 \\
P &= 36 \text{ W}
\end{aligned}
$$

Practice Problems 1-EE
Answers at End of Chapter

Using the power formula $P = I^2R$, calculate the power in watts for the following values of I in amperes and R in ohms.

1. $I = 2$, $R = 5$
2. $I = 5$, $R = 2$
3. $I = 0.1$, $R = 100$
4. $I = 10$, $R = 10$
5. $I = \sqrt{2}$, $R = 10$
6. $I = 0.02$, $R = 68$
7. $I = 2$, $R = 47$
8. $I = 47$, $R = 2$
9. $I = 3$, $R = 3$
10. $I = 6$, $R = 3$

Another application is calculating the percentage of error in making laboratory measurements. Suppose that you have a standard resistance R rated at 100 Ω. When you measure R with an ohmmeter, the value is 101 Ω.

What is the percentage of error? The following formula can be used:

$$\% \text{ Error} = \left(\frac{R_M - R_R}{R_R} \right) \times 100 \qquad (1\text{-}1)$$

where R_M is the measured value and R_R is the rated value. Multiplying by 100 converts the fraction to a percentage. For R_R of 100 and R_M of 101, then,

$$
\begin{aligned}
\% \text{ Error} &= \left(\frac{101 - 100}{100} \right) \times 100 \\
&= \frac{1}{100} \times 100
\end{aligned}
$$

$$\% \text{ Error} = 1\%$$

If R_M is less than R_R, the answer will be negative but the sign does not matter in calculating percentage of error.

Practice Problems 1-FF
Answers at End of Chapter

Using Formula (1-1), find the percentage error for the following values of rated resistance R_R and measured resistance R_M:

1. $R_R = 10$, $R_M = 10.1$
2. $R_R = 100$, $R_M = 99$
3. $R_R = 1000$, $R_M = 1050$
4. $R_R = 1000$, $R_M = 950$
5. $R_R = 10,000$, $R_M = 9500$
6. $R_R = 100,000$, $R_M = 95,000$
7. $R_R = 1,000,000$, $R_M = 1,800,000$
8. $R_R = 5,000,000$, $R_M = 6,000,000$

Review Problems
Answers to Odd-Numbered Problems at Back of Book

These problems summarize the arithmetic operations explained in this chapter.

1. $2.432 + 0.415 =$
2. $0.93 + 2.24 =$
3. $8.496 - 2.301 =$
4. $7.621 - 0.2 =$
5. $2.25 - 1.9 =$
6. $12.4 \times 2 =$
7. $12.4 \times 22 =$
8. $7.2 \times 1000 =$
9. $7200 \div 1000 =$
10. $475 \div 2.5 =$
11. $9^2 =$
12. $90^2 =$

13. $17^2 =$

14. $\sqrt{625} =$

15. $\sqrt{2416} =$

16. $(4)^2 + (2 + 3)^2 =$

17. $9 \div 2 + 6 =$

18. For $I = \sqrt{P/R}$, find I with $P = 50$ and $R = 2$

19. Find the average

20. Find the rms value for 0, 5, 7, and 10

21. Give 0.08 as a percentage

22. Convert 2% to a decimal fraction.

value for 0, 5, 7, and 10

Answers to Practice Problems

1-A										
	1. Two	**7.** Three			**7.** 8567			**3.** 0.052		
	2. Three	**8.** Four			**8.** 556			**4.** 0.048		
	3. Four	**9.** One	1-G	**1.** 0.43				**5.** 1624.83		
	4. Two	**10.** Two		**2.** 0.072				**6.** 0.08		
	5. Five	**11.** Three		**3.** 0.59				**7.** 1075		
	6. Eight	**12.** Four		**4.** 0.0004				**8.** 775		
1-B	**1.** 30	**7.** 50		**5.** 0.0022				**9.** 26.45		
	2. 0	**8.** 0		**6.** 0.77				**10.** 30.2		
	3. 90	**9.** 0		**7.** 0.887				**11.** 234		
	4. 20	**10.** 50		**8.** 0.995				**12.** 5,999,711.4		
	5. 60	**11.** 50	1-H	**1.** 123		1-L	**1.** 9.6		**13.** 32	
	6. 50	**12.** 50		**2.** 99			**2.** 9.6		**14.** 32	
1-C	**1.** 25.	**7.** 606.		**3.** 100			**3.** 8.4			
	2. 225.	**8.** 7070.		**4.** 139			**4.** 27.06			
	3. 555.	**9.** 4.		**5.** 2555			**5.** 80			
	4. 505.	**10.** 40.		**6.** 3155			**6.** 80			
	5. 2003.	**11.** 400.		**7.** 10,432			**7.** 99.9			
	6. 8888.	**12.** 4000.		**8.** 21,293			**8.** 924			
1-D	**1.** 0.6	**9.** 0.06	1-I	**1.** 699			**9.** 56.28			
	2. 0.62	**10.** 0.006		**2.** 852			**10.** 54.6			
	3. 0.007	**11.** 0.02		**3.** 1459			**11.** 168			
	4. 7.3	**12.** 0.42		**4.** 795.2			**12.** 64			
	5. 0.205			**5.** 0.657		1-M	**1.** 15		**9.** 14	
	6. 0.999			**6.** 1.599			**2.** 12		**10.** 15	
	7. 0.000007			**7.** 569.32			**3.** 11			
	8. 0.200007			**8.** 259			**4.** 14			
1-E	**1.** 3 tenths	**9.** 1 tenth		**9.** 69.33			**5.** 17			
	2. 2 tenths	**10.** 4 tenths		**10.** 5,462,839			**6.** 206			
	3. 1 tenth	**11.** 5 tenths		**11.** 12,462,832			**7.** 19			
	4. 0 tenths	**12.** 5 tenths		**12.** 4632.843			**8.** 20			
	5. 3 tenths		1-J	**1.** 31		1-N	**1.** 1.4		**9.** 92	
	6. 2 tenths			**2.** 3.1			**2.** 1.2		**10.** 46	
	7. 1 tenth			**3.** 78.14			**3.** 1.1			
	8. 0 tenths			**4.** 88.22			**4.** 2.06			
1-F	**1.** 21			**5.** 111			**5.** 140			
	2. 321			**6.** 574			**6.** 120			
	3. 765			**7.** 0.0032			**7.** 12			
	4. 1555			**8.** 3,432,261			**8.** 2060			
	5. 977		1-K	**1.** 315		1-O	**1.** 33			
	6. 783			**2.** 285			**2.** 3.3			

3. 0.33		**13.** 7	**1-AA**	**1.** 1280
4. 33		**14.** 6		**2.** 1300
5. 7		**15.** 5		**3.** 1280
6. 0.7		**16.** 4		**4.** 0.128
7. 35.2		**17.** 3		**5.** 500
8. 372.18		**18.** 2		**6.** 490
9. 0.67		**19.** 1		**7.** 480
10. 0.33		**20.** 0		**8.** 0.333

11. 262.47

12. 8.19

1-V
1. 196 **9.** 54
2. 529 **10.** 392
3. 2916 **11.** 25
4. 153,664 **12.** 100
5. 361
6. 625
7. 14
8. 23

1-BB
1. No
2. Yes
3. Yes
4. No
5. No
6. Yes
7. No
8. Yes

1-P
1. 700 **9.** 850
2. 782 **10.** 850
3. 7820
4. 78,200
5. 482
6. 48.2
7. 9640
8. 96,400

1-W
1. 10 **9.** 8.5
2. 83.75 **10.** 7.17
3. 0.6
4. 1
5. 4.25
6. 5.75
7. 63.7 V (but 74.3 if 0 reading omitted)
8. 314.5 Ω (but 86.0 if 1000 reading omitted)

1-CC
1. 10 V **9.** 50 V
2. 10 V **10.** 70 V
3. 9 V
4. 0.3 V
5. 470 V
6. 0.02 V
7. 12,000 V
8. 14.14 V

1-Q
1. 0.7 **9.** 53.2
2. 7 **10.** 53.2
3. 7
4. 0.07
5. 2.34
6. 0.234
7. 89.6
8. 0.4

1-X
1. 5.2 **9.** 3.3
2. 7.98 **10.** 4.3
3. 2.7 **11.** 6.48
4. 31.1
5. 5.07
6. 0.507
7. 6.7
8. 5.34

1-DD
1. 2 A **9.** 0.012 A
2. 5 A **10.** 0.006 A
3. 0.003 A
4. 0.003 A
5. 10 A
6. 0.000 002 A
7. 0.003 A
8. 1.414 A

1-R
1. 2
2. 5.8
3. 5.432
4. 57
5. 0.0057
6. 0.0794
7. 3.6
8. 0.00036
9. 3790
10. 37.9

1-Y
1. 16
2. 7
3. 21
4. 59
5. 26
6. 13
7. 2
8. 20

1-EE
1. 20 W **9.** 27 W
2. 50 W **10.** 108 W
3. 1 W
4. 1000 W
5. 20 W
6. 0.0272 W
7. 188 W
8. 4418 W

1-S See Practice Problems 1-T
1-T See Practice Problems 1-S
1-U
1. 81
2. 64
3. 49
4. 36
5. 25
6. 16
7. 9
8. 4
9. 1
10. 0
11. 9
12. 8

1-Z
1. 12 **8.** 30
2. 8 **9.** 14
3. 7 **10.** 15
4. 0
5. 72
6. 29
7. 9

1-FF
1. 1%
2. 1%
3. 5%
4. 5%
5. 5%
6. 5%
7. 80%
8. 20%

2 NEGATIVE NUMBERS

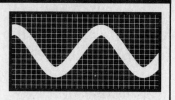

Is it possible for a physical quantity to be less than absolute zero? The answer is no. Even a small decimal fraction like 0.000 001 is more than zero, although it is only equal to one millionth of something. However, in practical calculations with physical quantities, it is often necessary to take into account the fact that the values may be related to a reference point. For instance, a force can be directed to the right or left of a reference point. Then a negative sign may be used for the number to show its direction from that reference point. The value of zero is used for the reference.

Negative numbers are explained in the following sections:

2-1 Positive and Negative Numbers
2-2 Addition of Negative Numbers
2-3 Subtraction of Negative Numbers
2-4 Multiplication and Division of Negative Numbers

2-1 Positive and Negative Numbers

The idea of positive and negative numbers in opposite directions is illustrated in Fig. 2-1. Positive values are usually considered to be to the right for the horizontal direction or upward for the vertical direction. Such values are called *signed numbers*. The + or − sign is an *operator* that indicates direction for the absolute value or magnitude. Usually the + sign is omitted. For instance, 8 is the same as +8, but a negative 8 must be shown as −8.

An important application of signed numbers is to label positive and negative voltage polarities. For instance, we can have 9 V or −9 V, with respect to the reference of 0 V. One is just as good as the other, but the opposite polarities produce opposite directions of current in an electric circuit.

2-2 Addition of Negative Numbers

To add a negative number to a positive number, take the difference between the two numbers and give the answer the sign of the larger number.

Example Add 8 and −5.

Answer $8 + (-5) = 3$
The 8 is the larger number. Therefore, the answer is positive. This problem is illustrated in Fig. 2-2 with arrows for the quantities in opposite directions. The shaded area shows 5 of the positive units canceled by 5 of the negative units, with the result being 3 positive units remaining. Adding a negative quantity is really the same as subtracting a positive quantity.

Example Add −8 and 5.

Answer $(-8) + 5 = -3$
Here we have a case where the larger number is negative. Therefore the answer is negative, or −3. This problem is illustrated in Fig. 2-3.

Example Add −8 and −3.

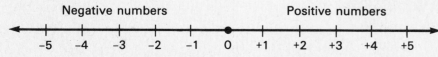

Fig. 2-1 Positive and negative values have opposite directions.

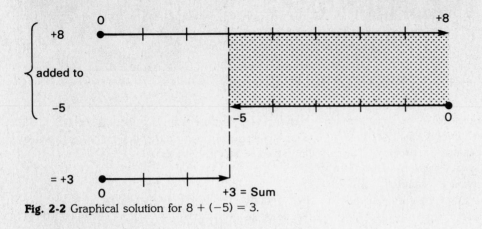

Fig. 2-2 Graphical solution for $8 + (-5) = 3$.

Answer $-8 + (-3) = -11$
The 8 and 3 total 11, but because both are negative the answer is -11.

The addition can be in any order. As an example, $8 + (-5) = 3$ is the same as $-5 + 8 = 3$.

Practice Problems 2-A
Answers at End of Chapter

Add these values.

1. $8 + 5 =$
2. $8 + (-5) =$
3. $(-5) + 8 =$
4. $5 + (-8) =$
5. $(-5) + (-8) =$
6. $(-8) + (-5) =$
7. $(-8) + 19 =$
8. $8 + (-15) =$
9. $28 + (-8) =$
10. $(-5) + (-3) =$
11. $(-3) + (-5) =$
12. $4 + (-9) =$

2-3 Subtraction of Negative Numbers

The rule is, Change the sign of the number to be subtracted (the *subtrahend*) and add. The addition is done by the rules just given in Sec. 2-2.

Example   Subtract -3 from 5. The -3 is the subtrahend.

Answer $5 - (-3) = 5 + 3 = 8$

Note that subtracting a negative number is the same as adding a positive number. The reason is that we have two reversals of direction in combining the values. The negative sign means the opposite direction of the value, and the subtraction is the reverse of addition.

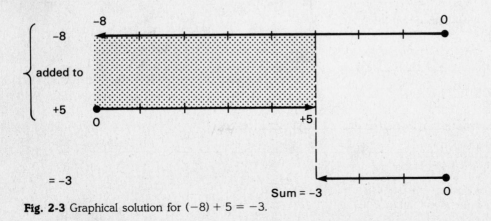

Fig. 2-3 Graphical solution for $(-8) + 5 = -3$.

Example Subtract -3 from -5. The -3 is the subtrahend.

Answer $-5 - (-3) = -5 + 3 = -2$
This problem is the same as adding -5 and 3.

Practice Problems 2-B
Answers at End of Chapter

Subtract these values.

1.	$8 - (+5) =$	6.	$8 - (-15) =$
2.	$8 - (-5) =$	7.	$19 - (-6) =$
3.	$(-5) - (+8) =$	8.	$(-12) - (-5) =$
4.	$(-5) - (-8) =$	9.	$72 - (-3) =$
5.	$8 - (+15) =$	10.	$80 - (+5) =$

Practice Problems 2-C
Answers at End of Chapter

Add these values.

1.	$8 + (-5) =$	6.	$8 + 15 =$
2.	$8 + 5 =$	7.	$19 + 6 =$
3.	$(-5) + (-8) =$	8.	$(-12) + 5 =$
4.	$(-5) + 8 =$	9.	$72 + (-3) =$
5.	$8 + (-15) =$	10.	$80 + (-5) =$

2-4 Multiplication and Division of Negative Numbers

When multiplying two numbers, if only one number is negative, the answer is negative.

Examples $6 \times (-3) = -18$
$(-6) \times 3 = -18$

For division also, when either of the two numbers is negative, the answer is negative.

Examples $(-6) \div 3 = -2$
$6 \div (-3) = -2$

The reason for the negative answer is that multiplication is merely a shortcut for addition, and adding a

series of negative numbers always results in a negative answer. Division is a shortcut for subtraction, and subtracting a series of negative numbers always results in a negative answer.

When both numbers in a multiplication or division problem are negative, the answer is positive.

Examples $(-6) \times (-3) = 18$
$(-6) \div (-2) = 3$

The reason for the positive answer is that after the series of negative numbers are added (as in multiplication) or subtracted (as in division), the negative answer is reversed and made positive. In short, two reversals of direction in the operations with two negative numbers bring the result back to the positive direction.

It is important to remember that the results of multiplying or dividing two negative numbers are different from the results of adding or subtracting them. When you add or subtract two negative numbers, the sum or difference is still negative. For multiplication or division with two negative numbers, though, the answer is positive.

In general, when a string of more than two numbers is multiplied, an even number of negative values means that the answer is positive. Then the pairs of negative numbers are combined for a positive value. As an example,

$$4 \times (-2) \times 6 \times (-3)$$
First $4 \times (-2) = -8$
then $-8 \times 6 = -48$
finally $-48 \times (-3) = 144$

The same rule applies to division. In summary, then, with an even number of negative values, the operations of multiplication and division result in a positive answer.

When an odd number of negative values is multiplied or divided, however, the answer is negative. This follows from the fact that the pairs of negative values can be combined for a positive value, but one negative value is always left to be multiplied or divided.

Example Show that the following is true:
$(-3) \times (-4) \times (-2) = -24$.

Answer $(-3) \times (-4) = 12$
$12 \times (-2) = -24$

The two negative numbers −3 and −4 are multiplied for the positive product of 12. Then this positive 12 is multiplied by −2 for the answer of −24.

All these rules for multiplication and division with negative numbers can be summarized as follows:

$$(-) \times (+) = (-)$$
$$(+) \times (-) = (-)$$
$$(-) \div (+) = (-)$$
$$(+) \div (-) = (-)$$
$$(-) \times (-) = (+)$$
$$(-) \div (-) = (+)$$

For any even number of negative values, that is, two, four, six, and so forth, the results of multiplication and division are positive.

An odd number of negative values, that is, one, three, five, etc., results in a negative answer.

Practice Problems 2-D
Answers at End of Chapter

Do the following multiplications for a positive or negative answer.

1. $4 \times (-3) =$
2. $(-4) \times 3 =$
3. $(-3) \times (-3) =$
4. $7 \times (-2) =$
5. $3 \times 3 \times (-3) =$
6. $(-2) \times (-2) \times 5 =$
7. $3 \times (-4) =$
8. $(-6) \times ½ =$
9. $(-3) \times (-2) \times (-1) =$
10. $(-3) \times 2 \times (-1) =$

Practice Problems 2-E
Answers at End of Chapter

Do the following divisions for a positive or negative answer.

1. $12 \div (-3) = -4$
2. $(-12) \div 3 = -4$
3. $(-12) \div (-3) = +4$
4. $8 \div (-4) = -2$
5. $12 \div 3 = 4$
6. $12 \div (-4) = -3$
7. $[8 \times (-4)] \times 2 = -64$
8. $4 \times \{[(-3) \times 6] \times (-2)\} = 144$
9. $(-6) \div (-2) \div (-1) =$
10. $(-6) \div 2 \div (-1) =$

All these rules for negative values will apply to decimal fractions, and also to numbers greater than 1.

Practice Problems 2-F
Answers at End of Chapter

Do the following multiplications and divisions for a positive or negative answer.

1. $4 \times 2 \times (-3) =$
2. $(-4) \times 3 \times (-3) =$
3. $(-4) \times 2 \div (-8) =$
4. $4 \times 2 \times (-1) =$
5. $(-8) \times 6 \div (-2) =$
6. $(-6) \times (-6) =$
7. $(-3) \times (-3) \times (-3) =$
8. $8 \times (-3) \times 1 =$
9. $(-6) \div (-2) \times (-1) =$
10. $(-6) \div (2) \times (-1) =$

Review Problems
Answers to Odd-Numbered Problems at Back of Book

The following problems summarize operations with negative numbers.

1. $(-7) + 2 = -5$
2. $7 - 2 = 5$
3. $7 - (-2) = 9$
4. $(-7) \times (-2) = 14$
5. $(-7) \div (-2) = 14$
6. $9 - 2 - 3 =$
7. $4 \times (-2) \times (-3) =$
8. $(-9) \times (-2) \div (-3) =$
9. $(-8) \div (4) =$
10. $4 \div (-8) =$
11. Compare the answers for Practice Problems 2-B and 2-C.
12. $(-0.12) + (0.12) =$
13. $(-0.12) - (+0.12) =$
14. $42 \times (-0.5) =$
15. $(-42) \div (2) =$

Answers to Practice Problems

2-A **1.** 13
    2. 3
    3. 3
    4. −3
    5. −13
    6. −13
    7. 11
    8. −7
    9. 20
    10. −8
    11. −8
    12. −5

2-B **1.** 3
    2. 13
    3. −13
    4. 3
    5. −7
    6. 23
    7. 25
    8. −7
    9. 75
    10. 75

2-C **1.** 3
    2. 13
    3. −13
    4. 3
    5. −7
    6. 23
    7. 25
    8. −7
    9. 69
    10. 75

2-D **1.** −12
    2. −12
    3. 9
    4. −14
    5. −27
    6. 20
    7. −12
    8. −3
    9. −6
    10. 6

2-E **1.** −4
    2. −4
    3. 4
    4. −2
    5. 4
    6. −3
    7. −64
    8. 144
    9. −3
    10. 3

2-F **1.** −24
    2. 36
    3. 1
    4. −8
    5. 24
    6. 36
    7. −27
    8. −24
    9. −3
    10. 3

3 FRACTIONS

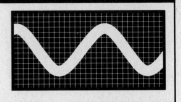

A fraction bar indicates the arithmetic operation of division. For example, ⅔ means 2 divided by 3. The 2 above the fraction bar is the *numerator,* while the 3 below is the *denominator*. The numerator is divided by the denominator. A fraction such as ⅔ is less than 1 because the numerator is less than the denominator. This type of fraction is called a *proper fraction* because the value is less than 1. A fraction like 4/3 is an *improper fraction* because the value is more than 1. Actually 4/3 is equal to 1⅓.

A proper fraction represents part of or less than one complete unit. For proper fractions, remember that as the denominator becomes larger, the value of the fraction becomes smaller, assuming the numerator remains the same. For instance, ¼ is less than ½.

Methods of working with proper fractions are explained in the following sections:

3-1 Multiplication of Fractions

The rule for the multiplication of fractions is, Multiply the numerators to obtain a new numerator and multiply the denominators to obtain a new denominator.

Example Multiply $\dfrac{2}{3}$ by $\dfrac{4}{7}$.

Answer $\dfrac{2}{3} \times \dfrac{4}{7} = \dfrac{2 \times 4}{3 \times 7} = \dfrac{8}{21}$

It may seem surprising that the product of two fractions less than 1 must be less than either of the two fractions. The reason is that this multiplication takes only a fractional part of the original fraction.

The fractions can be multiplied in any order.

Example $\dfrac{2}{5} \times \dfrac{1}{3} = \dfrac{2}{15}$

or $\dfrac{1}{3} \times \dfrac{2}{5} = \dfrac{2}{15}$

Practice Problems 3-A
Answers at End of Chapter

Multiply the fractions.

1. $\dfrac{1}{3} \times \dfrac{2}{3} =$ 4. $\dfrac{3}{4} \times \dfrac{3}{7} =$

2. $\dfrac{2}{7} \times \dfrac{1}{3} =$ 5. $\dfrac{3}{5} \times \dfrac{2}{9} =$

3. $\dfrac{4}{9} \times \dfrac{2}{3} =$ 6. $\dfrac{4}{7} \times \dfrac{2}{3} =$

3-2 Division of Fractions

The rule for the division of fractions is, invert the fraction that is the divisor and then multiply.

Example Divide $\dfrac{2}{5}$ by $\dfrac{5}{7}$.

Answer Write this as

$\dfrac{2}{5} \div \dfrac{5}{7}$

The divisor 5/7 is inverted to 7/5 and then multiplied by the other number.

$\dfrac{2}{5} \times \dfrac{7}{5} = \dfrac{14}{25}$

In division we must keep the fractions in the proper order. Thus ⅖ divided by 5/7 is not the same as 5/7 divided by ⅖. To prove this, we can do the problems both ways.

$$\frac{2}{5} \div \frac{5}{7} = \frac{2}{5} \times \frac{7}{5} = \frac{14}{25}$$

In the second case,

$$\frac{5}{7} \div \frac{2}{5} = \frac{5}{7} \times \frac{5}{2} = \frac{25}{14}$$

which is the inverse of the correct answer.

Practice Problems 3-B
Answers at End of Chapter

Divide the fractions.

1. $\dfrac{1}{3} \div \dfrac{4}{5} =$ 3. $\dfrac{3}{8} \div \dfrac{8}{9} =$

2. $\dfrac{3}{7} \div \dfrac{2}{3} =$ 4. $\dfrac{3}{5} \div \dfrac{5}{7} =$

3-3 The Simplest Form of a Fraction

A fraction is usually easiest to work with when the numerator and denominator have their lowest possible values. Consider, for instance, the fractions 50/100, 12/24, 3/6, and 125/250. Without actually realizing why, we may know that each of these numbers is equal to ½. If we were going to use them in further calculations, it would be considerably easier to work with ½ than with 125/250; yet the end result would be exactly the same.

Example	Multiply $\dfrac{2}{9}$ by $\dfrac{125}{250}$.
Answer	$\dfrac{2}{9} \times \dfrac{125}{250} = \dfrac{250}{2250} = \dfrac{1}{9}$
But	$\dfrac{2}{9} \times \dfrac{1}{2} = \dfrac{2}{18} = \dfrac{1}{9}$

To reduce a fraction to simpler numbers, the numerator and denominator can be *divided* by the same number. This does not change the actual value of the fraction; it simply states the fraction in numbers that are easier to handle. In the fraction 250/2250, both numerator and denominator were divided by 250; in the fraction 2/18, both were divided by 2.

Practice Problems 3-C
Answers at End of Chapter

Reduce to the lowest possible numerator and denominator.

1. $\dfrac{6}{9} =$ 3. $\dfrac{5}{20} =$

2. $\dfrac{6}{12} =$ 4. $\dfrac{9}{21} =$

Addition and subtraction of fractions often require finding a common denominator. This may involve *multiplying* numerator and denominator by the same number.

Example	Change ⅔ to a fraction with a denominator of 18.
Answer	If we multiply the denominator by 6, then the result will be 18. But we must also multiply the numerator by the same number to keep the value of the fraction constant. This gives us $2 \times 6 = 12$ for the new numerator. Or
	$\dfrac{2}{3} \times \dfrac{6}{6} = \dfrac{12}{18}$

The fraction 6/6 is, of course, equal to 1, and multiplying by 1 does not change the original number.

This process of raising the denominator is the opposite of reducing a fraction to its simplest form, but both methods are used for different reasons. In multiplication or division of fractions, the simplest form is easier to work with because of smaller numbers. Also, in the final answer to a problem, the simplest form is better.

For addition and subtraction of fractions, however, it may be necessary to raise a denominator because all the denominators must be the same.

Practice Problems 3-D
Answers at End of Chapter

Raise each pair of fractions to the lowest common denominator.

1. $\dfrac{1}{2}$ and $\dfrac{1}{6}$ 3. $\dfrac{2}{5}$ and $\dfrac{3}{10}$

2. $\dfrac{1}{7}$ and $\dfrac{3}{14}$ 4. $\dfrac{2}{3}$ and $\dfrac{1}{6}$

Operations with fractions less than 1 can result in an answer of more than 1. For example, ⅔ + ⅔ is equal to 4/3. Such a value as 4/3 is an improper fraction. It can be converted to a mixed number, consisting of a whole number and a proper fraction. The method is to divide the numerator by the denominator. Then the answer (called the *quotient*) is a whole number with a remainder that is a proper fraction.

Example   Express the improper fraction 5/3 as a mixed number.

Answer $\dfrac{5}{3} = 5 \div 3 = 1\dfrac{2}{3}$

Practice Problems 3-E
Answers at End of Chapter

Change each improper fraction to a mixed number.

1. $\dfrac{7}{6} =$ 3. $\dfrac{6}{5} =$

2. $\dfrac{14}{3} =$ 4. $\dfrac{9}{4} =$

3-4 Multiplying or Dividing a Fraction by a Whole Number

A whole number can be thought of as an improper fraction. For instance, 4 is the same as 4/1.

Example   Multiply $\dfrac{2}{3}$ by 4.

Answer $\dfrac{2}{3} \times 4 = \dfrac{2}{3} \times \dfrac{4}{1} = \dfrac{8}{3}$

Reduced to a mixed number, the product 8/3 is equal to 2⅔. Note that the procedure here is the same as multiplying only the numerator of the fraction by the whole number.

For division, the fractional divisor is inverted to multiply.

Example   Divide $\dfrac{2}{3}$ by 4.

Answer $\dfrac{2}{3} \div 4 = \dfrac{2}{3} \div \dfrac{4}{1}$

$= \dfrac{2}{3} \times \dfrac{1}{4} = \dfrac{2}{12}$ or $\dfrac{1}{6}$

Note that this method of dividing is the same as multiplying only the denominator of the proper fraction by the whole number.

Practice Problems 3-F
Answers at End of Chapter

Multiply or divide. Reduce each answer to lowest terms.

1. $\dfrac{1}{6} \times 3 =$ 4. $\dfrac{1}{3} \div 3 =$

2. $\dfrac{3}{5} \times 5 =$ 5. $\dfrac{3}{5} \div 5 =$

3. $\dfrac{2}{7} \times 3 =$ 6. $\dfrac{2}{7} \div 3 =$

3-5 Addition and Subtraction of Fractions

Fractions must have the same denominator before they can be added or subtracted. Add or subtract the numerators of the fractions and put the result over the common denominator.

Example   Add the following fractions:

$$\frac{3}{7} + \frac{2}{7}$$

Answer $\frac{3}{7} + \frac{2}{7} = \frac{3+2}{7} = \frac{5}{7}$

Example   Subtract the following fractions as shown:

Answer $\frac{3}{7} - \frac{2}{7} = \frac{3-2}{7} = \frac{1}{7}$

If the denominators of the fractions are not the same, they must be changed before the fractions can be added or subtracted. To make the calculations as simple as possible, the *lowest common denominator* should be used.

Example   Add $\frac{5}{12}$ and $\frac{7}{18}$.

Answer   The example looks like this:

$$\frac{5}{12} + \frac{7}{18}$$

Since the denominators are not the same, the addition cannot be performed in this form. The multiples of 12 are 12, 24, 36, 48, etc. The multiples of 18 are 18, 36, 54, 72, etc. Since 36 is the lowest multiple common to both denominators, each will be changed to 36:

$$\frac{5 \times 3}{12 \times 3} + \frac{7 \times 2}{18 \times 2}$$

or

$$\frac{15}{36} + \frac{14}{36}$$

When the denominators are the same, the numerators are simply added and the result is put over the denominator 36:

$$\frac{15 + 14}{36} = \frac{29}{36}$$

Therefore,

$$\frac{5}{12} + \frac{7}{18} = \frac{29}{36}$$

Practice Problems 3-G
Answers at End of Chapter

Add or subtract the fractions.

1. $\frac{4}{9} + \frac{1}{9} =$ 4. $\frac{4}{9} - \frac{2}{9} =$

2. $\frac{5}{9} + \frac{4}{9} =$ 5. $\frac{5}{9} - \frac{4}{9} =$

3. $\frac{7}{9} + \frac{5}{9} =$ 6. $\frac{7}{9} - \frac{5}{9} + \frac{3}{9} =$

Practice Problems 3-H
Answers at End of Chapter

Combine the following fractions:

1. $\frac{4}{9} + \frac{1}{3} =$ 6. $2 + \left(4 \times \frac{1}{2}\right) =$

2. $\frac{4}{9} - \frac{1}{3} =$ 7. $\left(\frac{3}{7} \times \frac{2}{7}\right) + \frac{5}{49} =$

3. $\frac{3}{9} + \frac{2}{18} =$ 8. $\frac{5}{6} - \frac{1}{3} - \frac{1}{2} =$

4. $\frac{3}{4} - \frac{2}{5} =$ 9. $\left(\frac{3}{5} \times 2\right) - \frac{3}{5} =$

5. $\frac{1}{7} + \left(3 \times \frac{2}{7}\right) =$ 10. $\left(\frac{1}{3} \times 3\right) - \frac{1}{7} =$

3-6 Negative Fractions

A minus sign in front of a fraction bar makes the entire fraction negative. Use such a fraction as a negative number. When carrying out calculations, the minus sign is usually considered part of the numerator.

Example Add $\left(-\dfrac{1}{5}\right)$ to $\dfrac{3}{5}$.

$$\frac{3}{5} + \left(-\frac{1}{5}\right) = \frac{3}{5} - \frac{1}{5} = \frac{2}{5}$$

Example Subtract $\left(-\dfrac{1}{5}\right)$ from $\dfrac{3}{5}$.

$$\frac{3}{5} - \left(-\frac{1}{5}\right) = \frac{3}{5} + \frac{1}{5} = \frac{4}{5}$$

For multiplication and division, remember that a minus sign in either the numerator or the denominator makes the fraction negative.

Examples $\dfrac{3}{5} \times \left(\dfrac{-1}{5}\right) = \dfrac{-3}{25} = -\dfrac{3}{25}$

$\dfrac{3}{5} \times \left(\dfrac{1}{-5}\right) = \dfrac{3}{-25} = -\dfrac{3}{25}$

However, a minus sign in both the numerator and the denominator means the fraction is really positive. The reason is that division of the two negative numbers in the numerator and denominator would result in a positive quotient.

Example $\dfrac{3}{5} \times \left(\dfrac{-1}{-5}\right) = \dfrac{3}{5} \times \dfrac{1}{5} = \dfrac{3}{25}$

Practice Problems 3-I
Answers at End of Chapter

Combine these fractions.

1. $\dfrac{4}{9} - \left(-\dfrac{2}{9}\right) =$ 3. $\dfrac{4}{9} \div \left(\dfrac{-2}{9}\right) =$

2. $\dfrac{4}{9} \times \left(\dfrac{-2}{9}\right) =$ 4. $\dfrac{4}{9} \times \left(\dfrac{-2}{-9}\right) =$

3-7 Reciprocals and Decimal Fractions

A reciprocal of any number is 1 divided by that number. For instance, the reciprocal of 7 is $\frac{1}{7}$. Furthermore, the reciprocal of $\frac{1}{7}$ is equal to 7. The reason is:

$$1 \div \frac{1}{7} = 1 \times \frac{7}{1} = 7$$

The reciprocals of the digits 2 to 9 often must be converted to decimal numbers. These values are:

$\frac{1}{2} = 0.5$	$\frac{1}{6} = 0.166\cdots$
$\frac{1}{3} = 0.333\cdots$	$\frac{1}{7} = 0.142857\cdots$
$\frac{1}{4} = 0.25$	$\frac{1}{8} = 0.125$
$\frac{1}{5} = 0.2$	$\frac{1}{9} = 0.111\cdots$

The decimal values for $\frac{1}{3}$, $\frac{1}{6}$, $\frac{1}{7}$, and $\frac{1}{9}$ are inexact and can be rounded off as explained in Chap. 1. The values for $\frac{1}{2}$, $\frac{1}{4}$, $\frac{1}{5}$, and $\frac{1}{8}$ are exact to the number of places shown.

The decimal equivalents for each of these fractions result from just dividing each denominator into the numerator 1. Note also that $\frac{1}{1} = 1$ and $\frac{0}{1} = 0$. However, the division $\frac{1}{0}$ is indeterminate. Because it is extremely large without limits, it is generally called infinity and given the symbol ∞.

A special reciprocal worth memorizing is $1/(2\pi)$. This factor is often used in electronics formulas. The value of the constant π rounded off to the nearest hundredth is 3.14, and $2\pi = 6.28$. The reciprocal is:

$$\frac{1}{2\pi} = \frac{1}{6.28} = 0.159$$

The factor π is often used in the analysis of circular motion. Specifically, π is the constant ratio of the circumference to the diameter, for any size circle.

Practice Problems 3-J
Answers at End of Chapter

Give the reciprocal as a decimal value.

1. 5	3. $\sqrt{25}$	5. π	7. $6 + 3$
2. 50	4. $\frac{1}{2}$	6. 2π	8. $0.25 + 0.75$

In Probs. 7 and 8 notice that you must add before taking the reciprocal of the sum.

3-8 Working with Decimal Fractions

A decimal fraction does not have a fraction bar. An example is 0.5, which is the same as $^5/_{10}$, or ½. For this reason decimal fractions are often easier to work with; they do not have numerators and denominators. However, you must keep track of the decimal point.

Example Add 0.5, 0.2, and 0.1.

Answer This problem can be shown as

$$\begin{array}{r} 0.5 \\ 0.2 \\ \underline{0.1} \\ 0.8 \end{array}$$

For addition or subtraction, the decimal points for all the numbers are in line.

Multiplication and division of decimal fractions is the same as with whole numbers.

Example Multiply 0.4×0.2.

Answer This problem can be shown as

$$\begin{array}{r} 0.4 \\ \times\ 0.2 \\ \hline 0.08 \end{array}$$

There are two decimal places in the product, because each of the two factors has one decimal place.

Example Divide 0.6 by 0.2.

Answer This problem can be shown as

$$0.2\overline{)0.6} = 2\overline{)6} = 3$$

The decimal points are moved to make the divisor a whole number.

Practice Problems 3-K
Answers at End of Chapter

Solve the following:

1. $0.2 + 0.3 + 0.4 =$
2. $0.2 + 0.3 - 0.1 =$
3. $0.2 + 0.05 =$
4. $0.2 \times 0.3 =$
5. $0.2 \times (-0.3) =$
6. $0.3 \times 0.02 =$

Review Problems
Answers to Odd-Numbered Problems at Back of Book

The following problems summarize operations with fractions.

1. $\frac{1}{2} \times \frac{1}{8} =$
2. $\frac{1}{2} \times \frac{1}{5+3} =$
3. $\frac{1}{2} \div \frac{1}{8} =$
4. $\frac{1}{3} + \frac{2}{3} =$
5. $\frac{1}{3} - \frac{2}{3} =$
6. $\frac{2}{3} - \frac{1}{3} =$
7. $\frac{1}{4} \times \left(\frac{-1}{2}\right) =$
8. $1 \div \frac{1}{9} =$
9. $\frac{1}{2\pi} \times 2 =$
10. $\frac{1}{4} + 0.33 =$

Answers to Practice Problems

3-A
1. $^2/_9$
2. $^2/_{21}$
3. $^8/_{27}$
4. $^9/_{28}$
5. $^6/_{45}$
6. $^8/_{21}$

3-B
1. $^5/_{12}$
2. $^9/_{14}$
3. $^{27}/_{64}$
4. $^{21}/_{25}$

3-C
1. $^2/_3$
2. $^1/_2$
3. $^1/_4$
4. $^3/_7$

3-D
1. $^3/_6$ and $^1/_6$
2. $^2/_{14}$ and $^3/_{14}$
3. $^4/_{10}$ and $^3/_{10}$
4. $^4/_6$ and $^1/_6$

3-E
1. $1^1/_6$
2. $4^2/_3$
3. $1^1/_5$

4. $2\frac{1}{4}$

3-F **1.** $\frac{1}{2}$
2. 3
3. $\frac{6}{7}$
4. $\frac{1}{9}$
5. $\frac{3}{25}$
6. $\frac{2}{21}$

3-G **1.** $\frac{5}{9}$
2. 1
3. $1\frac{1}{3}$
4. $\frac{2}{9}$
5. $\frac{1}{9}$
6. $\frac{5}{9}$

3-H **1.** $\frac{7}{9}$
2. $\frac{1}{9}$
3. $\frac{4}{9}$
4. $\frac{7}{20}$
5. 1
6. 4
7. $\frac{11}{49}$
8. 0
9. $\frac{3}{5}$
10. $\frac{6}{7}$

3-I **1.** $\frac{6}{9}$ or $\frac{2}{3}$
2. $-\frac{8}{81}$
3. $-\frac{36}{18}$ or -2
4. $\frac{8}{81}$

3-J **1.** 0.2
2. 0.02
3. 0.2
4. 2
5. 0.318
6. 0.159
7. 0.111
8. 1

3-K **1.** 0.9
2. 0.4
3. 0.25
4. 0.06
5. -0.06
6. 0.006

4 POWERS AND ROOTS

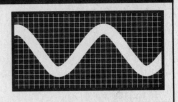

Raising a number to a higher power is the same as multiplying a number by itself a given number of times. The number of times the given number is used as a factor in multiplication is called the power. The power is indicated by an exponent written above and to the right of the given number. For instance, 5×5 is written as 5^2. Either way, the answer is 25. The 5 is the *base* number here, raised to the second power as shown by the exponent 2. The second power is called the square of the number. In other words, 5 squared is equal to 25. The cube is the third power of a number. Then $5 \times 5 \times 5 = 5^3$ or 125. Any base number can be raised to any power.

More details are explained in the following sections:

4-1   Positive Exponents
4-2   Roots of Positive Numbers
4-3   Squares and Roots for the Digits
4-4   Powers of a Negative Number
4-5   Roots of a Negative Number
4-6   Powers and Roots of Fractions
4-7   Powers and Roots of Numbers with Exponents
4-8   Square and Roots with Factors
4-9   Squares and Roots with Terms
4-10   Using a Calculator for Powers and Roots

4-1   Positive Exponents

The purpose of using positive exponents is to provide a shortcut method of indicating repeated multiplication of the same number.

Examples	$2 \times 2 \times 2 = 2^3 = 8$
	$5 \times 5 \times 5 = 5^3 = 125$
	$10 \times 10 \times 10 = 10^3 = 1000$

In the last example, note that 10^3 saves the time and space of multiplying 10 three times. Similarly 10^6 is equal to 1,000,000. The exponents are especially useful for powers of 10, as explained in Chap. 5.

Practice Problems 4-A
Answers at End of Chapter

Raise the base number to the power indicated by the exponent and give the answer.

1.	$2^3 = 8$	**7.**	$2^4 = 16$	**13.**	$10^5 =$
2.	$3^2 = 9$	**8.**	$7^2 = 49$	**14.**	$10^6 =$
3.	$3^3 = 27$	**9.**	$7^3 =$	**15.**	$5^2 =$
4.	$6^2 = 36$	**10.**	$10^2 =$	**16.**	$5^3 =$
5.	$8^2 = 64$	**11.**	$10^3 =$	**17.**	$16^2 =$
6.	$8^3 = 512$	**12.**	$10^4 =$	**18.**	$16^3 =$

4-2   Roots of Positive Numbers

The *root* of a number is the value that can be multiplied by itself to equal the original number.

Example	Find the cube root of 125. This can be written as
	$$\sqrt[3]{125}$$
Answer	Although there are arithmetic procedures that can be used to find roots, most often either a table of roots or an electronic calculator with this capability is used. For certain commonly used numbers, the square roots and cube roots are memorized. For the above example we find
	$$5 \times 5 \times 5 = 125$$
	Therefore,
	$$\sqrt[3]{125} = 5$$
Examples	Find each of the following roots.
	$\sqrt{4} = 2$ [and -2, since $(-2) \times (-2) = 4$]
	$\sqrt[3]{8} = 2$ [but *not* -2, since $(-2) \times (-2)$ $\times (-2) = -8$]
	$\sqrt[4]{16} = 2$ (Again, -2 is also an answer.)
	$\sqrt[5]{32} = 2$

Practice Problems 4-B
Answers at End of Chapter

Find the value of the indicated root.

1.	$\sqrt[3]{8} =$	**5.**	$\sqrt{36} =$	**9.**	$\sqrt{25} =$
2.	$\sqrt{9} =$	**6.**	$\sqrt[3]{64} =$	**10.**	$\sqrt[3]{27} =$
3.	$\sqrt[4]{10,000} =$	**7.**	$\sqrt{64} =$	**11.**	$\sqrt{49} =$
4.	$\sqrt[3]{125} =$	**8.**	$\sqrt[4]{16} =$	**12.**	$\sqrt{81} =$

The radical sign $\sqrt{}$ is generally used with an index number to indicate the root. When no index number is shown, it is assumed to be 2 for the square root.

A root can also be indicated as a fractional exponent. As examples, $64^{1/2}$ is the same as $\sqrt{64}$ and $8^{1/3}$ is the same as $\sqrt[3]{8}$. The general form for this conversion is

$$\sqrt[x]{y} = y^{1/x}$$

Let y be 9 and x equal 2. Then

$$\sqrt[x]{y} = \sqrt[2]{9} = 9^{1/2} = 3$$

4-3 Squares and Roots for the Digits

Because they occur so often in numerical problems, the squares of the digits should be memorized. These are as follows:

$$0^2 = 0 \times 0 = 0 \qquad 5^2 = 5 \times 5 = 25$$
$$1^2 = 1 \times 1 = 1 \qquad 6^2 = 6 \times 6 = 36$$
$$2^2 = 2 \times 2 = 4 \qquad 7^2 = 7 \times 7 = 49$$
$$3^2 = 3 \times 3 = 9 \qquad 8^2 = 8 \times 8 = 64$$
$$4^2 = 4 \times 4 = 16 \qquad 9^2 = 9 \times 9 = 81$$

It should be noted that any power or root of zero is still zero. Also, any power or root of 1 is still 1.

The square of a number is that number multiplied by itself. It does *not* mean double the number. It happens that $2^2 = 4$, which is also double the 2, but this is true only with the digit 2. The square of 3 is 9, but doubling 3 would give 6.

Practice Problems 4-C
Answers at End of Chapter

Give the square of the following.

1.	$5^2 =$	**4.**	$7^2 =$	**7.**	$12^2 =$
2.	$3^2 =$	**5.**	$10^2 =$	**8.**	$4^2 =$
3.	$1^2 =$	**6.**	$9^2 =$	**9.**	$8^2 =$

The squares of the digits should be learned forward and backward for some important square roots also. Since $5^2 = 25$, then the square root of 25 is 5, as $5 \times 5 = 25$. Some common square root values that should be memorized are the following:

$$\sqrt{1} = 1 \qquad\qquad \sqrt{25} = 5$$
$$\sqrt{2} = 1.414 \qquad \sqrt{36} = 6$$
$$\sqrt{3} = 1.732 \qquad \sqrt{49} = 7$$
$$\sqrt{4} = 2 \qquad\qquad \sqrt{64} = 8$$
$$\sqrt{9} = 3 \qquad\qquad \sqrt{81} = 9$$
$$\sqrt{16} = 4 \qquad\quad \sqrt{100} = 10$$

The reason why the squares of the digits and their roots are used so often is the fact that very large or small numbers can be converted to these values as a factor with the appropriate multiple of 10.

Example Express 900 as a factor with a multiple of 10.

Answer $900 = 9 \times 100 = 9 \times 10^2$

Practice Problems 4-D
Answers at End of Chapter

Give the following square roots. Check your answer by multiplying.

1.	$\sqrt{25} =$	**4.**	$\sqrt{100} =$	**7.**	$\sqrt{16} =$
2.	$\sqrt{36} =$	**5.**	$\sqrt{9} =$	**8.**	$\sqrt{49} =$
3.	$\sqrt{81} =$	**6.**	$\sqrt{4} =$	**9.**	$\sqrt{144} =$

4-4 Powers of a Negative Number

Since raising a number to a power is the same as repeated multiplication, the rules for multiplying negative numbers apply here.

Raising a negative number to an even power, for example, squaring a negative number, results in a positive answer.

Examples $(-3)^2 = (-3) \times (-3) = 9$
$(-2)^4 = (-2) \times (-2) \times (-2) \times (-2)$
$= 16$

Raising a negative number to an odd power, for example, cubing a negative number, results in a negative answer.

Examples $(-2)^3 = (-2) \times (-2) \times (-2) = -8$
$(-3)^5 = (-3) \times (-3) \times (-3) \times (-3)$
$\times (-3)$
$= -243$

Practice Problems 4-E
Answers at End of Chapter

Raise these numbers to the power indicated.

1. $(3)^2 =$	**5.** $(-1)^6 =$
2. $(-3)^2 =$	**6.** $(+1)^6 =$
3. $(2)^3 =$	**7.** $(-4)^2 =$
4. $(-1)^5 =$	**8.** $(-4)^3 =$

4-5 Roots of a Negative Number

Since an even power of a negative number always leads to a positive number answer, it is not possible to work back from a negative number to a positive even root. However, it is possible to find odd roots of negative numbers.

Example   Find the cube root of -8.

Answer $\sqrt[3]{-8} = -2$
To prove the answer we need only find the *cube* of -2.
$(-2)^3 = (-2) \times (-2) \times (-2)$
$= -8$

Though we cannot calculate the square root of a negative number by arithmetic, such numbers are useful in electricity and electronics. To get around this problem, we *assume* that square roots do exist for negative numbers. These roots are called *imaginary numbers*. Since the positive and negative numbers we have been dealing with up to this point can be represented on the horizontal axis, the imaginary numbers are indicated on a vertical axis, called the *j axis*. In order to indicate that a number is imaginary, it is preceded by the letter *j*, representing $\sqrt{-1}$.

Example   Represent $\sqrt{-4}$ as an imaginary number.

Answer $\sqrt{-4} = \sqrt{-1} \times \sqrt{4} = j\sqrt{4} = j2$

Since *j* preceding a number indicates that the number is to be rotated to the vertical *j* axis, we often use the expression *j operator*.* The *j* operator really represents the angle of 90°.

Practice Problems 4-F
Answers at End of Chapter

Find the indicated root.

1. $\sqrt[3]{8} =$	**5.** $\sqrt{-64} =$
2. $\sqrt[3]{-8} =$	**6.** $\sqrt[3]{-64} =$
3. $\sqrt[3]{-27} =$	**7.** $\sqrt[3]{125} =$
4. $\sqrt{25} =$	**8.** $\sqrt[4]{16} =$

Another possibility is to have a number that is positive but has a negative exponent, such as 10^{-2}. The negative exponent, however, is only a method of indicating a reciprocal. For instance, 5^{-2} is the reciprocal of 5^2. And 3^{-2} is the reciprocal of 3^2. These values can be stated as follows:

$$5^2 = 25 \qquad\qquad 3^2 = 9$$
$$5^{-2} = \frac{1}{5^2} = \frac{1}{25} \quad \text{and} \quad 3^{-2} = \frac{1}{3^2} = \frac{1}{9}$$

also

$$10^2 = 100$$
$$10^{-2} = \frac{1}{10^2} = \frac{1}{100} = 0.01$$

Note that these reciprocal values are positive numbers. The negative value is only in the exponent. The negative exponents for reciprocals are especially useful with powers of 10 to indicate decimal fractions. This application is explained in detail in the next chapter.

4-6 Powers and Roots of Fractions

When a fraction is raised to a power, both the numerator and the denominator must be raised to that power.

---

*For details of the *j* operator, see Bernard Grob, *Basic Electronics*, Chap. 25, Macmillan/McGraw-Hill School Publishing Company.

Example Cube ⅔.

Answer The numerator must be cubed, and the denominator must also be cubed.

$$\left(\frac{2}{3}\right)^3 = \frac{2 \times 2 \times 2}{3 \times 3 \times 3} = \frac{8}{27}$$

To find the root of a fraction the reverse process is used in that the root of both the numerator and the denominator must be found.

Example Find $\sqrt[3]{8/27}$.

Answer $\sqrt[3]{\dfrac{8}{27}} = \dfrac{\sqrt[3]{8}}{\sqrt[3]{27}} = \dfrac{2}{3}$

This answer, when cubed, will result in the original fraction.

Practice Problems 4-G
Answers at End of Chapter

Find the powers or roots of the following fractions.

1. $(⅔)^2 =$ 5. $\sqrt{4/9} =$
2. $(3/7)^2 =$ 6. $\sqrt{9/49} =$
3. $(½)^4 =$ 7. $\sqrt{1/16} =$
4. $(⅓)^2 =$ 8. $\sqrt{1/9} =$

An interesting fact about a proper fraction is that raising it to a power makes the fraction smaller. For instance, $(½)^2 = ¼$. The answer of ¼ is smaller than ½ because the denominator is larger with the same numerator. Remember that a proper fraction has a value less than 1 because the denominator is larger than the numerator. Raising the fraction to a power accentuates this property by increasing the larger number in the denominator more than the increase in the numerator.

Practice Problems 4-H
Answers at End of Chapter

Pick out the larger value in each of the following pairs of fractions.

1. ½ or ¼   5. 1/100 or 1/10
2. ⅔ or 4/9   6. 10/1 or 1/10
3. ⅕ or 1/25   7. 0.1 or 0.01
4. 9/49 or 3/7   8. 0.5 or 0.25

4-7 Powers and Roots of Numbers with Exponents

Very often, it is necessary to raise to a power or find the root of a number that already has an exponent, as in $(2^2)^3$. To perform this operation the rule is, multiply the exponents and make the product the new exponent for the original base number.

Example Find the value of $(4^2)^3$.

Answer The product of the exponents is

$$2 \times 3 = 6$$

This 6 is the new exponent for the base 4. Then the answer is

$$4^6$$

The value of 4^6 is equal to

$$4 \times 4 \times 4 \times 4 \times 4 \times 4 = 4096$$

This value is the same as $(16)^3$:

$$16 \times 16 \times 16 = 4096$$

The 16 is equal to 4^2.

A similar but opposite process is used for finding roots. The rule is, divide the original exponent by the root and use the answer as the new exponent of the original base.

Example Find $\sqrt[3]{8^6}$.

Answer Divide the exponent 6 by the root 3. The result is

$$\frac{6}{3} = 2$$

The 2 is the new exponent for the original base 8. Thus

$$\sqrt[3]{8^6} = 8^2 = 64$$

Example　Find $\sqrt[6]{8^2}$.

Answer　Divide the exponent 2 by the root 6. The result is

$$\frac{2}{6} = \frac{1}{3}$$

The fraction $\frac{1}{3}$ is the new exponent for the original base 8. Thus

$$8^{1/3} = \sqrt[3]{8} = 2$$

Practice Problems 4-I
Answers at End of Chapter

Find the powers and roots, as indicated.

1. $(9^2)^2 =$
2. $[(-3)^2]^3 =$
3. $\sqrt{7^2} =$
4. $(4 \times 2^2)^2 =$
5. $\sqrt[4]{8^4} =$
6. $\sqrt[3]{10^6} =$
7. $\sqrt[3]{5^6} =$
8. $\sqrt{36 \times 8^4} =$

4-8 Squares and Roots with Factors

Factors are parts of numbers which, when multiplied together, produce the number. In $2 \times 4 = 8$, the 2 and 4 are factors of 8.

In squaring or taking the square root, the operation can be applied to each of the factors separately.

Examples　Find the following square roots, using factors:

$$\sqrt{25 \times 9} = \sqrt{25} \times \sqrt{9} = 5 \times 3 = 15$$
$$\sqrt{49 \times 4} = \sqrt{49} \times \sqrt{4} = 7 \times 2 = 14$$
$$\sqrt{49 \times 25} = \sqrt{49} \times \sqrt{25} = 7 \times 5 = 35$$

Examples　Find the following squares, using factors:

$$(2 \times 3)^2 = (2)^2 \times (3)^2 = 4 \times 9 = 36$$
$$(4 \times 5)^2 = 4^2 \times 5^2 = 16 \times 25 = 400$$

This procedure of separating the factors can be used for any power or root, but the examples here are for the common problem of finding a square or square root.

Practice Problems 4-J
Answers at End of Chapter

Find the square or square root of the following.

1. $(4 \times 2)^2 =$
2. $(9 \times 3^2)^2 =$
3. $(5^2 \times 3^2)^2 =$
4. $(7 \times 10^4)^2 =$
5. $\sqrt{16 \times 4} =$
6. $\sqrt{81^4} =$
7. $\sqrt{2^4 \times 3^2} =$
8. $\sqrt{49 \times 10^8} =$ *22/359.93*

4-9 Squares and Roots with Terms

Terms are numbers in a group that are to be added or subtracted. For instance, in $(2 + 7)$, the 2 and 7 are terms. To find the square or root, *all the terms must be combined first*.

Example　Square $(2 + 7)$, or find $(2 + 7)^2$.

Answer　This procedure is different from the method with factors. If you square each term separately, the answer will be wrong. First combine the terms by adding: $2 + 7 = 9$. Then square the sum: $(9)^2 = 81$. Therefore, the answer is $(2 + 7)^2 = 81$.

Note that $(2 + 7)^2$ is not equal to $2^2 + 7^2$, as $4 + 49 = 53$.

Practice Problems 4-K
Answers at End of Chapter

Solve the following.

1. $(3 + 4)^2 =$
2. $[(3 \times 10^3) + (4 \times 10^3)]^2 =$
3. $(3 - 4)^2 =$
4. $[(2 \times 10^{-4}) + (3 \times 10^{-4})]^2 =$
5. $(2 + 4)^2 =$
6. $(3 + 7 + 9)^2 =$
7. $(5 + 2)^2 =$
8. $\sqrt{(42 + 7)} =$

The same procedure applies to the square root for a group of terms. They must all be combined before the root is found. For instance, $\sqrt{16 + 9} = \sqrt{25} = 5$. If you take the roots of each number separately, the answer of $4 + 3 = 7$ will be wrong. A power or root for factors can be applied separately, but terms 4 and 3 must be combined first.

Example	Find $\sqrt{72 - 8}$.
Answer	Combine terms first: $72 - 8 = 64$. Then find the root $\sqrt{64} = 8$.

This procedure of combining terms before you find the square or square root also applies for any power or root. Always keep in mind that different rules are used for terms that are added or subtracted and factors that are multiplied or divided.

Practice Problems 4-L
Answers at End of Chapter

Solve the following.

1. $\sqrt{16 + 9} =$
2. $\sqrt{35 + 1} =$
3. $\sqrt{5^2 - 4^2} =$
4. $\sqrt{(8 \times 10^6) + 10^6} =$
5. $\sqrt{5^2 - 3^2} =$
6. $\sqrt{12 + 7 + 6} =$
7. $\sqrt{9 + 7} =$
8. $\sqrt{3^2 + 16} =$

As an application of these methods, an important formula in electronics is $Z = \sqrt{R^2 + X^2}$, where Z is the impedance, R the resistance, and X the reactance in the circuit, all in ohms units. To calculate Z, the R and X within the square root sign must be evaluated and combined before you take the square root of the grouping.

Example	If $R = 3\ \Omega$ and $X = 4\ \Omega$, find Z.
Answer	$Z = \sqrt{R^2 + X^2} = \sqrt{3^2 + 4^2}$ $= \sqrt{9 + 16} = \sqrt{25}$ $Z = 5\ \Omega$

Practice Problems 4-M
Answers at End of Chapter

With the formula $Z = \sqrt{R^2 + X^2}$, find Z for the following values of R and X.

1. $R = 3, X = 4$
2. $R = 4, X = 3$
3. $R = 4, X = 4$
4. $R = 3, X = -4$
5. $R = 6, X = -6$
6. $R = 4, X = 8$

Practice Problems 4-N
Answers at End of Chapter

With the formula $X = \sqrt{Z^2 - R^2}$, find X for the following values of Z and R.

1. $Z = 5, R = 4$
2. $Z = 14.14, R = 10$
3. $Z = 8.48, R = 6$
4. $Z = 6, R = 2$
5. $Z = 8, R = 4$
6. $Z = 17, R = 9$

4-10 Using a Calculator for Powers and Roots

The scientific type of electronic calculator usually has a key for raising almost any number to any power. This key is generally labeled $\boxed{y^x}$, where y is the number and x is the exponent. To take an example:

Example	Using a calculator, raise 2 to the third power, that is, find the value of 2^3.
Answer	Using $\boxed{y^x}$, then y is 2 and x is 3. The procedure is

1. Punch in $\boxed{2}$ on the keyboard for the number.
2. Press $\boxed{y^x}$.
3. Punch in $\boxed{3}$ for the exponent.
4. Press $\boxed{=}$ to display the answer of 8.

The proof is $2 \times 2 \times 2 = 8$.

As another example, find the value of 7^4. Punch in $\boxed{7}$ for the base number, press $\boxed{y^x}$, punch in $\boxed{4}$ for the exponent, and press $\boxed{=}$ for the answer of 2401. The proof is $7 \times 7 \times 7 \times 7 = 2401$.

Practice Problems 4-O
Answers at End of Chapter

Find each of the following using the $\boxed{y^x}$ key.

1. $2^4 =$
2. $2^5 =$
3. $2^6 =$
4. $2^8 =$
5. $8^3 =$
6. $8^4 =$
7. $16^2 =$
8. $16^3 =$
9. $16^4 =$
10. $(49)^5 =$
11. $(6)^6 =$
12. $(0.1)^6 =$

For the reverse procedure of finding the root of a number, some calculators have a key labeled $\sqrt[x]{y}$. Here y is the number and x is the index number for the root. For example, to find the cube root of 8, or $\sqrt[3]{8}$,

1. Punch in 8 on the keyboard for the number.
2. Press $\sqrt[x]{y}$.
3. Punch in 3 for the root.
4. Press $=$ to display the answer of 2.

The proof is that $2 \times 2 \times 2 = 8$.

On many calculators, the exponent and root functions are on the same key, but you must press another key first to choose which function you want. The reversal key is sometimes labeled $2nd$ (for second function), INV (for inverse), or F (for function).

As another example, find the fourth root of 2401 or $\sqrt[4]{2401}$. Punch in 2 4 0 1 for the number, press $\sqrt[x]{y}$, punch in 4 for the fourth root, and press $=$ for the answer of 7. The proof is $7 \times 7 \times 7 \times 7 = 2401$.

Practice Problems 4-P
Answers at End of Chapter

Find each of the following using the $\sqrt[x]{y}$ key.

1. $\sqrt[3]{8} =$
2. $\sqrt[6]{64} =$
3. $\sqrt[3]{4096} =$
4. $\sqrt[4]{10,000} =$
5. $\sqrt[7]{869} =$
6. $\sqrt[2]{1,000,000} =$
7. $\sqrt[2]{4692} =$
8. $\sqrt[3]{674} =$
9. $\sqrt[5]{98} =$
10. $\sqrt[3]{100} =$
11. $\sqrt[3]{27,000} =$
12. $\sqrt[15]{87,000,000} =$

The calculator cannot solve for the power or root of a negative number because of the way the calculator handles numbers internally. Keep in mind the following:

1. With a negative base number, an even power results in a positive answer but an odd power results in a negative answer. For example, $(-2)^2 = 4$ but $(-2)^3 = -8$.
2. With a positive number, an even root is both positive and negative. For example, $\sqrt{4}$ is $+2$ and -2.

It should be noted, though, that the calculator can handle a negative exponent, as it is just the reciprocal of the positive exponent with the same base. The general form of this conversion is

$$x^{-n} = \frac{1}{x^n}$$

As an example, let x be 2 and the exponent be 3. Then

$$x^{-n} = 2^{-3}$$
$$= \frac{1}{2^3}$$
$$= \frac{1}{8}$$
$$= 0.125$$

When using the calculator for a negative exponent, just punch it in and press the sign reversal $\pm$ key. You should see the negative sign on display. Note that the reversal key is not the subtraction $-$ key. Then proceed as described before. As an example, for 2^{-3} the steps are:

1. Punch in 2 for the base number y.
2. Press y^x.
3. Punch in 3 for the exponent x.
4. Press the $\pm$ key to reverse the sign to -3.
5. Press the $=$ key to display the answer of 0.125.

Review Problems
Answers to Odd-Numbered Problems at Back of Book

The following problems summarize operations with powers and roots.

1. $3^4 =$ 81
2. $\sqrt[3]{27} =$ 3
3. $(3.7)^2 =$ 13.69
4. $\sqrt{13.69} =$ 3.7
5. $\sqrt{64} =$ 8
6. $(5)^3 =$ 125
7. $(-4)^3 =$ -64
8. $\sqrt[3]{-64} =$ -4
9. $(-5)^4 =$
10. $\sqrt[4]{625} =$ 5
11. $(½)^3 =$ .125 3×10^{-4}
12. $\sqrt{⅛} =$ .35
13. $(0.25)^2 =$
14. $\sqrt{0.0625} =$ .25
15. $(3 \times 10^4)^2 =$
16. $\sqrt{9 \times 10^8} =$ 30800
17. $(2 + 5)^2 =$
18. $(6 - 4)^3 =$ 8
19. $(2 + 3)^2 \times 4 =$
20. $(6 - 4)^3 - 5 =$ 3

Answers to Practice Problems

4-A	1.	8	13.	100,000
	2.	9	14.	1,000,000
	3.	27	15.	25
	4.	36	16.	125
	5.	64	17.	256
	6.	512	18.	4096
	7.	16		
	8.	49		
	9.	343		
	10.	100		
	11.	1000		
	12.	10,000		

4-B	1.	2	9.	5
	2.	3	10.	3
	3.	10	11.	7
	4.	5	12.	9
	5.	6		
	6.	4		
	7.	8		
	8.	2		

4-C	1.	25	7.	144
	2.	9	8.	16
	3.	1	9.	64
	4.	49		
	5.	100		
	6.	81		

4-D	1.	5	7.	4
	2.	6	8.	7
	3.	9	9.	12
	4.	10		
	5.	3		
	6.	2		

4-E	1.	9
	2.	9
	3.	8
	4.	−1
	5.	+1
	6.	1
	7.	16
	8.	−64

| 4-F | 1. | 2 |
| | 2. | −2 |

	3.	−3
	4.	5
	5.	$j8$
	6.	−4
	7.	5
	8.	2

4-G	1.	$4/9$
	2.	$9/49$
	3.	$1/16$
	4.	$1/9$
	5.	$2/3$
	6.	$3/7$
	7.	$1/2$
	8.	$1/3$

4-H	1.	$1/2$
	2.	$2/3$
	3.	$1/5$
	4.	$3/7$
	5.	$1/10$
	6.	$10/1$
	7.	0.1
	8.	0.5

4-I	1.	6561
	2.	729
	3.	7
	4.	256
	5.	8
	6.	100
	7.	25
	8.	384

4-J	1.	64
	2.	6561
	3.	50,625
	4.	49×10^8
	5.	8
	6.	6561
	7.	12
	8.	7×10^4

4-K	1.	49	7.	49
	2.	49×10^6	8.	7
	3.	1		
	4.	25×10^{-8}		
	5.	36		
	6.	361		

4-L	1.	5
	2.	6
	3.	3
	4.	3×10^3
	5.	4
	6.	5
	7.	4
	8.	5

4-M	1.	5
	2.	5
	3.	5.657
	4.	5
	5.	8.485
	6.	8.944

4-N	1.	3
	2.	10
	3.	6
	4.	5.7
	5.	6.93
	6.	14.4

4-O	1.	16
	2.	32
	3.	64
	4.	256
	5.	512
	6.	4096
	7.	256
	8.	4096
	9.	65,536
	10.	2.82×10^8
	11.	46,656
	12.	0.000 001

4-P	1.	2
	2.	2
	3.	16
	4.	10
	5.	2.63
	6.	1000
	7.	68.5
	8.	8.77
	9.	2.5
	10.	4.64
	11.	30
	12.	3.38

5 POWERS OF 10

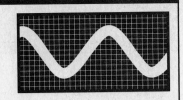

From Chap. 4 we learned that a power or exponent is written above and to the right of a number to indicate how many times the number is used as a factor in multiplication by itself. As an example, 10^3 is the same as $10 \times 10 \times 10$, which equals 1000. The exponent here is 3, and 10 is the *base* for the exponent. The base is raised to a power indicated by the exponent.

The base 10 is common for exponents, because 10 is the basis of decimal numbers for counting and for the decimal multiples in the metric system of units. (See Chap. 7.) In general, powers of 10 help keep track of the decimal point in arithmetic operations involving very large or very small numbers.

More details are explained in the following sections:

5-1 Positive Exponents of 10

Numbers greater than 1 can be written as powers of 10 by using positive exponents.

Note that 10 and 10^1 are the same. A number written without a power is assumed to have the exponent 1.

A higher positive exponent means a larger number. For instance, 10^3 for 1000 is more than 10^2 for 100.

Example   Represent 1, 10, 100, 1000, 10,000, 100,000, and 1,000,000 using powers of 10.

Answer

Power of 10	Multiplication	Product
10^0	—	1
10^1	10	10
10^2	10×10	100
10^3	$10 \times 10 \times 10$	1000
10^4	$10 \times 10 \times 10 \times 10$	10,000
10^5	$10 \times 10 \times 10 \times 10 \times 10$	100,000
10^6	$10 \times 10 \times 10 \times 10 \times 10 \times 10$	1,000,000

Practice Problems 5-A
Answers at End of Chapter

Convert to powers of 10.

1. $100 = 10^2$
2. $1000 = 10^3$
3. $10,000 = 10^4$
4. $100,000 = 10^5$
5. $1,000,000 = 10^6$
6. $10,000,000 = 10^7$

Practice Problems 5-B
Answers at End of Chapter

Write the following as common numbers.

1. $10^3 = 1000$
2. $10^2 = 100$
3. $10^6 = 1,000,000$
4. $10^4 = 10000$

5-2 Negative Exponents of 10

Numbers less than 1 can also be written as powers of 10. In this case a negative exponent must be used. A negative exponent of 10 shows powers of tenths, compared to 10 for a positive exponent.

Example   Write 0.1, 0.01, 0.001, 0.0001, 0.000 01, and 0.000 001 using powers of 10.

Answer

Power of 10	Multiplication	Product
10^{-1}	$\frac{1}{10}$	0.1
10^{-2}	$\frac{1}{10} \times \frac{1}{10}$	0.01
10^{-3}	$\frac{1}{10} \times \frac{1}{10} \times \frac{1}{10}$	0.001
10^{-4}	$\frac{1}{10} \times \frac{1}{10} \times \frac{1}{10} \times \frac{1}{10}$	0.0001
10^{-5}	$\frac{1}{10} \times \frac{1}{10} \times \frac{1}{10} \times \frac{1}{10} \times \frac{1}{10}$	0.00001
10^{-6}	$\frac{1}{10} \times \frac{1}{10} \times \frac{1}{10} \times \frac{1}{10} \times \frac{1}{10} \times \frac{1}{10}$	0.000001

Practice Problems 5-C
Answers at End of Chapter

Convert to powers of 10.

1. $0.001 = 10^{-3}$ 4. $100 = 10^{2}$ 7. $\frac{1}{100} = 10^{-2}$

2. $\frac{1}{1000} = 10^{-3}$ 5. $\frac{1}{1,000,000} = 10^{-6}$ 8. $0.0001 = 10^{-4}$

3. $0.01 = 10^{-2}$ 6. $0.000\ 001 = 10^{-6}$

Practice Problems 5-D
Answers at End of Chapter

Convert to decimal fractions.

1. $10^{-3} = 0.001$ 3. $10^{-2} = 0.01$

2. $10^{-1} = 0.1$ 4. $10^{-6} = 0.000001$

Practice Problems 5-E
Answers at End of Chapter

Convert to proper fractions.

1. $10^{-3} = \frac{1}{1000}$ 3. $10^{-2} = \frac{1}{100}$

2. $10^{-1} = \frac{1}{10}$ 4. $10^{-6} = \frac{1}{1000000}$

Notice that the larger the negative exponent, the smaller the number. For instance, 10^{-3} is smaller than 10^{-2}, as $\frac{1}{1000}$ is less than $\frac{1}{100}$. When you increase the negative exponent by 1, the number becomes 10 times smaller. To go the other way, decrease a negative exponent by 1 to make the number 10 times larger. These rules for negative exponents are opposite from the rules for positive exponents.

Practice Problems 5-F
Answers at End of Chapter

Change the exponent to make the following numbers 10 times larger.

1. 10^{3} 10^{4} 5. 10^{7} 10^{8}
2. 10^{-3} 10^{-2} 6. 10^{-7} 10^{-6}
3. 10^{-6} 10^{-5} 7. 10^{-5} 10^{-4}
4. 10^{6} 10^{7} 8. 10^{5} 10^{6}

Practice Problems 5-G
Answers at End of Chapter

Change the exponent to make the following numbers 10 times smaller.

1. 10^{3} 10^{2} 5. 10^{7} 10^{6}
2. 10^{-3} 10^{-4} 6. 10^{-7} 10^{-8}
3. 10^{-6} 10^{-7} 7. 10^{-5} 10^{-6}
4. 10^{6} 10^{5} 8. 10^{5} 10^{4}

5-3 Converting to Powers of 10

All the numbers used so far in this chapter have been perfect powers of 10. However, use of the powers of 10 is not limited to these numbers. The general procedure for using powers of 10 with any number is to convert the given number into two factors, where one factor (the power of 10) is used merely to place the decimal point.

Example	Write 750 as the product of two numbers, one of which is a power of 10.
Answer	$75 \times 10 = 750$
	$7.5 \times 10^{2} = 750$
	$0.75 \times 10^{3} = 750$

Notice that the power of 10 merely positions the decimal point in the original number. The exponent equals the number of places the decimal point is moved. What we have, then, is another way of writing a number so that the original number is easier to handle. In other words, we have a shorthand way of writing otherwise long numbers.

Example Write 1,640,000 using powers of 10 for millions.

Answer $1,640,000 = 1.64 \times 1,000,000$
$= 1.64 \times 10^6$

Example Write each of the following as a number between 1 and 10 multiplied by a power of 10: 100; 980; 9800; 9840.

Answer $100 = 1 \times 100 = 1 \times 10^2$
$980 = 9.8 \times 100 = 9.8 \times 10^2$
$9800 = 9.8 \times 1000 = 9.8 \times 10^3$
$9840 = 9.84 \times 1000 = 9.84 \times 10^3$

In the above examples only positive exponents were used, because the numbers are greater than 1. However, the same procedure applies to decimal fractions less than 1. In this case, negative exponents are used to mark off the number of places the decimal point is moved for multiples of tenths.

Example Write each of the following as a number between 1 and 10 multiplied by a power of 10: 0.01; 0.05; 0.053.

Answer $0.01 = 1 \times 0.01 = 1 \times 10^{-2}$
$0.05 = 5 \times 0.01 = 5 \times 10^{-2}$
$0.053 = 5.3 \times 0.01 = 5.3 \times 10^{-2}$

Moving the point to the right multiplies the value for a coefficient greater than 1. However, the negative exponent is a division for the same number of places, to keep the number the same. In short, 0.05 and 5.0×10^{-2} are two ways to write the same value.

The Coefficient of the Base In a number like 9×10^3, the 9 is a coefficient of base 10 with its power. The coefficient is a factor to be multiplied. In this example the coefficient 9 and the base number 10^3, or 1000, are multiplied for the product $9 \times 1000 = 9000$.

When no coefficient is given, it is assumed to be 1. For instance, 1000 can be written as 1×10^3. Also, 0.01 is 1×10^{-2}.

The purpose of having the coefficient is to factor out the part of the number that is not a perfect power of 10. In this way all numbers can be written as powers of 10 with the appropriate coefficient.

Scientific or Engineering Notation Numbers that are written in the form of a power of 10 and a coefficient of 1 or more, but less than 10, are said to be written in scientific notation. For instance, 4.5×10^4 is in this notation. In other words, the coefficient has only one decimal place to the left of the decimal point.

However, engineering notation is usually preferable in electrical and electronics work. In engineering notation the powers of 10 are always given in multiples of 3, that is, 10^3, 10^6, 10^9, 10^{12}, 10^{-3}, 10^{-6}, 10^{-9}, 10^{-12}. This is because components and values are usually specified or stated in terms of metric prefixes such as milli (10^{-3}), micro (10^{-6}), mega (10^6), nano (10^{-9}), as examples. When this is done, the coefficient is written to accommodate the power of 10. The metric system is explained in Chap. 7.

Example Write 15,000 in both scientific and engineering terms.

Answer Scientific notation:

$15,000 = 1.5 \times 10^4$

Engineering notation:

$15,000 = 15 \times 10^3$

Practice Problems 5-H
Answers at End of Chapter

Convert to powers of 10, in both scientific and engineering notation.

1. $400 = $ _0.4×10³_
2. $470 = $ _0.47×10³_
3. $4000 = $ _0.4×10³_
4. $4700 = $ _4.7×10³_
5. $8,000,000 = $ _8×10⁶_
6. $0.04 = $ _40×10³_
7. $0.047 = $ _4.7×10³_
8. $0.004 = $ _4×10⁻³_
9. $0.0047 = $ _4.7×10⁻³_
10. $0.000\ 008 = $ _8×10⁻⁶_

Practice Problems 5-I
Answers at End of Chapter

Convert each of the following to a number without the power of 10 notation.

1. $4 \times 10^2 = 400$
2. $4.7 \times 10^2 = 470$
3. $4 \times 10^3 = 4000$
4. $4.7 \times 10^3 = 4700$
5. $8 \times 10^6 = 8,000\,000$
6. $4 \times 10^{-2} = 0.04$
7. $4.7 \times 10^{-2} = 0.047$
8. $4 \times 10^{-3} = 0.004$
9. $4.7 \times 10^{-3} = 0.0047$
10. $8 \times 10^{-6} = 0.000008$

5-4 Multiplication with Powers of 10

To multiply numbers made up of powers of 10, multiply the coefficients to obtain the new coefficient and add the exponents to obtain the new power of 10.

Example	Multiply 200 by 40,000.
Answer	$200 = 2 \times 10^2$
	$40,000 = 4 \times 10^4$

Then the problem is

$$(2 \times 10^2) \times (4 \times 10^4)$$

$2 \times 4 = 8$	for the coefficients
$2 + 4 = 6$	for the exponents

Finally,

$$(2 \times 10^2) \times (4 \times 10^4)$$
$$= 8 \times 10^6$$

or

$$200 \times 40,000 = 8,000,000$$

The reason why the exponents are added for multiplication of the base is that each increase of 1 in a positive exponent is equivalent to moving the decimal point one place, as when multiplying by 10.

Practice Problems 5-J
Answers at End of Chapter

Multiply in powers of 10.

1. $(2 \times 10^2) \times (3 \times 10^4) = 6 \times 10^6$
2. $(4 \times 10) \times (2 \times 10) = 8 \times 10^2 = 80$
3. $(7 \times 10^7) \times (1 \times 10) = 7 \times 10^8$
4. $(3 \times 10^7) \times (2 \times 10^8) = 6 \times 10^{15}$
5. $(5 \times 10^5) \times (1 \times 10^2) = 5 \times 10^7$
6. $(3 \times 10^3) \times (2 \times 10^2) = 6 \times 10^5$
7. $(7 \times 10^2) \times (1 \times 10^2) = 7 \times 10^4$
8. $(2.5 \times 10^4) \times (2 \times 10^4) = 5 \times 10^8$
9. $(2 \times 10^4) \times (3 \times 10^2) = 6 \times 10^6$
10. $(3 \times 10^2) \times (2 \times 10^3) = 6 \times 10^5$

When there are negative exponents, they are also added. The new exponent is a larger negative number.

Example	Multiply 0.02 by 0.1.
Answer	$0.02 = 2 \times 10^{-2}$
	$0.1 = 1 \times 10^{-1}$

Then the problem is

$$(2 \times 10^{-2}) \times (1 \times 10^{-1})$$

$1 \times 2 = 2$	for the coefficients
$(-2) + (-1) = -3$	for the exponents

Finally,

$$(2 \times 10^{-2}) \times (1 \times 10^{-1})$$
$$= 2 \times 10^{-3}$$

or

$$0.02 \times 0.1 = 0.002$$

Note that the coefficient 2 in 2×10^{-3} is still positive. Only the exponent -3 is negative, indicating the fraction $\frac{1}{1000}$ for 10^{-3}. However, the value of the complete number 2×10^{-3} is still positive. It is a positive fraction, less than 1 because of the negative exponent.

Practice Problems 5-K
Answers at End of Chapter

Multiply in powers of 10.

1. $(2 \times 10^{-2}) \times (3 \times 10^{-4}) =$ *6 × 10⁻⁶*
2. $(4 \times 10^{-1}) \times (2 \times 10^{-1}) =$ *8 × 10⁻²*
3. $(7 \times 10^{-7}) \times (1 \times 10^{-1}) =$ *7 × 10⁻⁸*
4. $(3 \times 10^{-7}) \times (2 \times 10^{-8}) =$ *6 × 10⁻¹⁵*
5. $(5 \times 10^{-5}) \times (1 \times 10^{-2}) =$ *5 × 10⁻⁷*
6. $(3 \times 10^{-3}) \times (2 \times 10^{-2}) =$ *6 × 10⁻⁵*
7. $(7 \times 10^{-2}) \times (1 \times 10^{-2}) =$ *7 × 10⁻⁴*
8. $(2.5 \times 10^{-4}) \times (2 \times 10^{-4}) =$ *5 × 10⁻⁸*
9. $(3 \times 10^{-4}) \times (3 \times 10^{-3}) =$ *9 × 10⁻⁷*
10. $(1 \times 10^{-1}) \times (1 \times 10^{-1}) =$ *1 × 10⁻²*

For the case of adding positive and negative exponents, take the difference between the two and give it the sign of the larger exponent.

Examples	Multiply the following:
	$(4 \times 10^{5}) \times (2 \times 10^{-3}) = 8 \times 10^{2}$
	$(4 \times 10^{-5}) \times (2 \times 10^{3}) = 8 \times 10^{-2}$

When there are more than two factors, just keep adding the exponents and multiplying the coefficients.

Example	Multiply the following:
	$(2 \times 10^{5}) \times (1 \times 10^{-4}) \times (3 \times 10^{3})$
Answer	$2 \times 1 \times 3 = 6$ for the coefficients
	$5 - 4 + 3 = 4$ for the exponents
	The final result is
	$(2 \times 10^{5}) \times (1 \times 10^{-4} \times (3 \times 10^{3})$
	$= 6 \times 10^{4}$

Practice Problems 5-L
Answers at End of Chapter

Do the following multiplications in powers of 10.

1. $(4 \times 10^{4}) \times (2 \times 10^{2}) =$ *8 × 10⁶*
2. $(5 \times 10^{3}) \times (3 \times 10^{3}) =$ *15 × 10⁶*
3. $(3 \times 10^{-2}) \times (2 \times 10^{-1}) =$ *6 × 10⁻³*
4. $(7 \times 10^{-1}) \times (4 \times 10^{-2}) =$ *28 × 10⁻³*
5. $(4 \times 10^{5}) \times (2 \times 10^{-3}) =$ *8 × 10²*
6. $(2 \times 10^{-5}) \times (3 \times 10^{3}) =$ *6 × 10⁻²*
7. $(3 \times 10^{7}) \times (2 \times 10^{2}) \times (2 \times 10^{-3}) =$ *12 × 10⁶*
8. $4,000,000 \times 2,000,000 =$ *8 × 10¹²*
9. $0.001 \times 0.003 =$ *3 × 10⁻⁶*
10. $4,000,000 \times 0.000\ 002 =$ *8 × 10*
11. $(7 \times 10^{4}) \times 200 =$ *14 × 10⁶*
12. $(4 \times 10^{-12}) \times (3 \times 10^{6}) =$ *12 × 10⁻⁶*

5-5 Division with Powers of 10

To divide numbers involving powers of 10, divide the coefficients but subtract the exponents.

Example	Divide 6,000,000 by 3000.
Answer	$6,000,000 = 6 \times 10^{6}$
	$3000 = 3 \times 10^{3}$
	Then the problem is
	$(6 \times 10^{6}) \div (3 \times 10^{3})$
	$6 \div 3 = 2$ for the coefficients
	$6 - 3 = 3$ for the exponents
	The final result is
	$(6 \times 10^{6}) \div (3 \times 10^{3}) = 2 \times 10^{3}$
	or
	$6,000,000 \div 3000 = 2000$

Only the powers of 10 are subtracted. The coefficients are still divided.

Remember that you must subtract the exponent for the divisor from the exponent for the dividend. If you do the reverse, the result is the reciprocal of the correct answer.

Practice Problems 5-M
Answers at End of Chapter

Divide in powers of 10.

1. $(8 \times 10^8) \div (2 \times 10^2) =$ *4 × 10⁶*
2. $(9 \times 10^9) \div (3 \times 10^3) =$ *3 × 10⁶*
3. $(7 \times 10^{18}) \div (2 \times 10^{15}) =$ *3.5 × 10³*
4. $(6 \times 10^6) \div (4 \times 10^4) =$ *1.5 × 10²*
5. $(5 \times 10^5) \div (2 \times 10^2) =$ *2.15 × 10³*
6. $(8 \times 8^8) \div (2 \times 10^2) =$ *4 × 10⁶*
7. $(7 \times 10^7) \div (1 \times 10^5) =$ *7 × 10²*
8. $(6 \times 10^{12}) \div (3 \times 10^8) =$ *2 × 10⁴*
9. $(6 \times 10^5) \div (3 \times 10^3) =$ *2 × 10²*
10. $(2 \times 10^4) \div (2 \times 10^2) =$ *1 × 10²*

When the divisor has a larger exponent, the subtraction results in a negative number for the new exponent.

Example   Divide 8×10^4 by 2×10^6.

Answer $(8 \times 10^4) \div (2 \times 10^6) = 4 \times 10^{-2}$

For the coefficients, we have

$$8 \div 2 = 4$$

For the exponents, we have

$$4 - 6 = -2$$

The answer of 4×10^{-2} is equal to 0.04.

When the divisor has a negative exponent which must be subtracted, just change its sign and add.

Example   Divide 6×10^5 by 2×10^{-3}.

Answer $(6 \times 10^5) \div (2 \times 10^{-3}) = 3 \times 10^8$

For the coefficients, we have

$$6 \div 2 = 3$$

For the exponents, we have

$$5 - (-3) = 5 + 3 = 8$$

The reason why the answer of 3×10^8 is larger than the original dividend of 6×10^5 is that the divisor is a decimal fraction less than 1.

For another possibility of division with negative powers of 10, both exponents can be negative.

Example   Divide 6×10^{-5} by 2×10^{-3}.

Answer $(6 \times 10^{-5}) \div (2 \times 10^{-3}) = 3 \times 10^{-2}$

For the coefficients, we have

$$6 \div 2 = 3$$

For the exponents, we have

$$-5 - (-3) = -5 + 3 = -2$$

Practice Problems 5-N
Answers at End of Chapter

Divide, using powers of 10.

1. $(1 \times 10^8) \div (1 \times 10^6) =$ *1 × 10²*
2. $(1 \times 10^6) \div (1 \times 10^8) =$ *1 × 10⁻²*
3. $(1 \times 10^6) \div (1 \times 10^{-8}) =$ *1 × 10¹⁴*
4. $(1 \times 10^{-6}) \div (1 \times 10^8) =$ *1 × 10⁻¹⁴*
5. $(1 \times 10^{-6}) \div (1 \times 10^{-8}) =$ *1 × 10²*
6. $60,000 \div 200 =$ *6 × 10⁴ ÷ 2 × 10² = 3 × 10²*
7. $60,000 \div 0.002 =$ *6 × 10⁴ ÷ 2 × 10⁻³ 3 × 10¹*
8. $0.000\ 016 \div 8 \times 10^3 =$ *16 × 10⁻⁴ ÷ 2 × 10⁻¹*
9. $(3 \times 10^3) \div (6 \times 10^5) =$ *5 × 10⁻²*
10. $(2 \times 10^2) \div (2 \times 10^4) =$ *1 × 10⁻²*

Earlier in the chapter the value of 10^0 was given as 1. The reason for this can be shown by a division problem, remembering that any number divided by itself is equal to 1, or unity.

Example   Using the rules of exponents, divide 1×10^3 by itself.

Answer $(1 \times 10^3) \div (1 \times 10^3) = 1 \times 10^0$

For the coefficients, $1 \div 1 = 1$.

For the exponents, $3 - 3 = 0$.

Since any number divided by itself must be equal to 1, we can say that

$$\frac{1 \times 10^3}{1 \times 10^3} = 1 \times 10^0 = 1$$

So 10^0 must be 1, as $1 \times 1 = 1$.

In fact, any number taken to the zero power must be 1.

Examples $5^0 = 1$
$8^0 = 1$
$\left(\dfrac{1}{2}\right)^0 = 1$
$(0.003)^0 = 1$

The general principle is that any number to the zero power corresponds to the fraction $\frac{1}{1}$, where the numerator and denominator are equal.

5-6 Reciprocals with Powers of 10

Just changing the sign of the exponent converts the base number to its reciprocal.

Example Find the reciprocal of 10^2.

Answer The reciprocal can be written as $1/10^2 = 10^{-2}$.
The values of $1/10^2$ and 10^{-2} are both equal to $\frac{1}{100}$, the reciprocal of 100.

Example Find the reciprocal of 10^{-2}.

Answer The reciprocal can be written as $1/10^{-2} = 10^2$.
Both values equal 100, the reciprocal of $\frac{1}{100}$.

Taking the reciprocal of a number is really a special case of division with the numerator of 1. Remember that 1 can be stated as 10^0. For the example of $1/10^2$, the division is

$$\frac{1}{10^2} = \frac{10^0}{10^2} = 10^{(0-2)} = 10^{-2}$$

In other words, with exponents the reciprocal means a change of sign, because the exponent is subtracted from a zero exponent. This rule applies to any base with any exponent. As another example, the reciprocal of 5^3 is equal to $1/5^3$ or 5^{-3}.

In these examples, the base actually has the coefficient of 1, but the reciprocal is still 1. With any other coefficient, you must find its reciprocal value besides changing the sign of the exponent for the reciprocal of the power of 10.

Example Find the reciprocal of 2×10^3.

Answer The reciprocal in this example can be written as $1/(2 \times 10^3)$.
Another way of writing this reciprocal is $\frac{1}{2} \times 1/10^3 = 0.5 \times 10^{-3}$.
The coefficient is 0.5, which is the reciprocal of 2, and 10^{-3}, which is the reciprocal of 10^3.

Practice Problems 5-O
Answers at End of Chapter

Give the following reciprocal values without the fraction bar, using powers of 10.

1. $\dfrac{1}{10^{-4}} =$ 10⁴ ✓

2. $\dfrac{1}{10,000} =$ 10⁻⁴

3. $\dfrac{1}{0.5 \times 10^{-2}} =$ 0.5×10²

4. $\dfrac{1}{0.25 \times 10^5} =$ 0.25×10⁻⁵ 4×10⁻⁵

5. $\dfrac{1}{100} =$ 10⁻²

6. $\dfrac{1}{2 \times 10^2} =$ 2×10⁻² 4×10⁻²

7. $\dfrac{1}{10^7} =$ 10⁻⁷

8. $\dfrac{1}{0.125 \times 10^{-8}} =$ 8×10⁸

5-7 Addition and Subtraction with Powers of 10

Only numbers having the same power of 10 can be added or subtracted directly. Add or subtract only the coefficients, keeping the same exponent. The reason for not changing the exponent is to leave the decimal place unchanged.

Example Add 6×10^3 and 2×10^3.

Answer $6 + 2 = 8$ for the coefficients

The power of 10 is not changed. Therefore,

$$(6 \times 10^3) + (2 \times 10^3) = 8 \times 10^3$$

Example Subtract 2×10^3 from 6×10^3.

Answer $6 - 2 = 4$ for the coefficients

The power of 10 is not changed. Therefore,

$$(6 \times 10^3) - (2 \times 10^3) = 4 \times 10^3$$

If the numbers do not have the same exponent, they must be changed so that the exponents are the same before you can add or subtract. Any exponent can be used, but they must all be the same.

Example Add 3×10^3 and 4×10^4.

Answer One of the exponents must be converted to equal the other before the addition can be done.
To change 10^4 to 10^3, you can divide the base by 10, but multiply the coefficient by 10 to keep the value of the number the same. For the base,

$$10^4 \div 10 = 10^3$$

For the coefficient,

$$4 \times 10 = 40$$

Then

$$4 \times 10^4 = 40 \times 10^3$$

Now both of the numbers in the example have the same exponent, and the addition can be carried through.

$$(3 \times 10^3) + (40 \times 10^3) = 43 \times 10^3$$

or

$$3000 + 40,000 = 43,000$$

It should be noted that using the same power of 10 corresponds to lining up the decimal points for addition or subtraction.

Practice Problems 5-P
Answers at End of Chapter

Add or subtract using powers of 10.

1. $(6 \times 10^3) + (2 \times 10^3) =$
2. $(6 \times 10^3) - (2 \times 10^3) =$
3. $(6 \times 10^{-3}) + (2 \times 10^{-3}) =$
4. $0.006 + 0.002 =$
5. $(6 \times 10^3) + (2 \times 10^4) =$
6. $6000 + 20,000 =$
7. $(4 \times 10^{-2}) + (5 \times 10^{-3}) =$
8. $0.04 + 0.005 =$
9. $0.001 + 0.002 =$
10. $(1 \times 10^{-3}) + (2 \times 10^{-3}) =$

5-8 Raising an Exponent of 10 to a Higher Power

To find the power of a number written with an exponent, raise the coefficient to the indicated power, but just multiply the exponents.

Example Raise 2×10^2 to the third power.

Answer This can be written as

$$(2 \times 10^2)^3$$

Raise the coefficient to the power indicated:

$$2^3 = 2 \times 2 \times 2 = 8$$

For the exponents,

$$2 \times 3 = 6$$

Therefore,

$$(2 \times 10^2)^3 = 8 \times 10^6$$

Remember that the coefficient 2 is a factor of the base 10. We can factor out the coefficient separately as

$$(2 \times 10^2)^3 = (2)^3 \times (10^2)^3$$
$$= 8 \times 10^6$$

Practice Problems 5-Q
Answers at End of Chapter

Raise to the power shown.

1. $(1 \times 10^2)^3 =$ *[1×10⁶]*   6. $(7 \times 10^3)^2 =$ *[49×10⁶]*
2. $(1 \times 10^3)^2 =$ *[1×10⁶]*   7. $(1 \times 10^7)^2 =$ *[1×10¹⁴]*
3. $(1 \times 10^6)^2 =$ *[1×10¹²]*   8. $(4 \times 10^{-6})^2 =$ *[16×10⁻¹²]*
4. $(1 \times 10^2)^7 =$ *[1×10¹⁴]*   9. $(2 \times 10^2)^2 =$ *[4×10⁴]*
5. $(1 \times 10^{-2})^3 =$ *[1×10⁻⁶]* 10. $(2 \times 10^2)^3 =$ *[8×10⁶]*

5-9 Taking a Root with Powers of 10

To find the root of a number written as a power of 10, take the root of the coefficient but just divide the exponent by the indicated root. This procedure is just the opposite of raising to a higher power.

Example Find the third root of 8×10^6.

Answer This can be written as

$$\sqrt[3]{8 \times 10^6} = \sqrt[3]{8} \times \sqrt[3]{10^6}$$

The third root of 8, also called the cube root, is found first:

$$\sqrt[3]{8} = 2 \text{ (since } 2 \times 2 \times 2 = 8)$$

For the exponents,

$$6 \div 3 = 2$$

Therefore,

$$\sqrt[3]{8 \times 10^6} = 2 \times 10^2$$

A root can also be expressed as a fractional power, using the reciprocal of the index number. The cube root is the same as the ⅓ power. The square root is the ½ power. For instance, $\sqrt{16}$ is the same as $16^{1/2}$, which equals 4. Also, $\sqrt[3]{8}$ or $8^{1/3}$ is equal to 2. A root as a fractional power follows the rules for multiplication and division of exponents.

A practical rule with roots is to have the exponent exactly divisible by the index number of the root. In other words, take square roots with an even-numbered exponent and cube roots with an exponent that is a

multiple of 3. Otherwise, taking the root results in a fractional exponent for base 10. For instance, $\sqrt{10^3} = 10^{3/2} = 10^{1.5}$, which has a fractional exponent. Such a number is not an exact multiple of ten. Its value can be determined, though, by logarithms or by using an electronic calculator. Usually, however, the calculations are easier with whole numbers for the exponents. You can always convert the exponent to the desired value before taking the root.

Example Find the square root of

$$40 \times 10^5$$

Answer This can be written as

$$\sqrt{40 \times 10^5}$$

Since the solution of this problem will involve dividing the exponent by 2, it would be better if the exponent were an even number rather than 5. The first step, then, is to convert 10^5 to either 10^4 or 10^6. To convert 10^5 to 10^6, merely multiply 10^5 by 10 and divide the coefficient by 10 so that the original number is not changed:

$$40 \times 10^5 \times \frac{10}{10} = \frac{40}{10} \times 10^5 \times 10$$
$$= 4 \times 10^6$$

Now the exponent is an even number, and the square root can be determined using the rule previously explained.

$$\sqrt{4 \times 10^6} = \sqrt{4} \times \sqrt{10^6}$$
$$= 2 \times 10^3$$

Practice Problems 5-R
Answers at End of Chapter

Find the indicated root.

1. $\sqrt{4 \times 10^8} =$ *[2×10⁴]*   6. $\sqrt{9 \times 10^{-6}} =$
2. $\sqrt[3]{27 \times 10^9} =$ 7. $\sqrt[5]{1 \times 10^{20}} =$ *[1×10⁴]*
3. $(10^{12})^{1/6} =$ 8. $(3.6 \times 10^7)^{1/2} =$
4. $\sqrt{1 \times 10^{10}} =$ 9. $\sqrt[3]{8 \times 10^9} =$ *[2×10³]*
5. $\sqrt[3]{1 \times 10^9} =$ *[1×10³]*   10. $(25 \times 10^8)^{1/2} =$

[handwritten: $4 \times 10^6 \times \frac{10}{10} = \frac{40}{10} \times 10^6 \times 10 = 4 \times 10^9 = 2 \times 10^3$]
[handwritten: 3×10^{-3}]
[handwritten: $1 \times 10^4 \cdot 1 \times 10^2$]

5-10 Summary of Arithmetic Operations with Digits 1 and 0

Working with 1 and 0 often can be confusing, because these digits have special properties.

1. For 1 as a factor:
    (a) $1 \times 1 = 1$. Multiplied any number of times, the answer is still 1.
    (b) A number multiplied by 1 remains the same. For instance, $10 \times 1 = 10$.
    (c) A number divided by 1 remains the same. For instance, $10 \div 1 = 10$.
2. For 0 as a factor:
    (a) The product of any number multiplied by 0 is 0.
    (b) Division by 0 is not a valid operation. In other words, in arithmetic, we cannot divide by 0.
3. For 1 as a base:
    (a) Any power or root of 1 is still 1. For instance, $1^3 = 1$, or $\sqrt{1} = 1$.
4. For 1 as an exponent:
    (a) Any base with exponent 1 has the same value as it has without the exponent. For instance, 10^1 is the same as 10.
5. For 0 as an exponent:
    (a) Any base with exponent 0 has the numerical value of 1. The reason is that the exponent 0 results from a base and exponent divided by the same base and exponent. Remember: dividing any number by itself results in the answer 1.

Practice Problems 5-S
Answers at End of Chapter

Do the indicated operations.

1. $10 \times 0 =$
2. $10 \times 1 =$
3. $(10)^0 \times 5 =$
4. $\sqrt[7]{1} =$
5. $14 \times 0 =$
6. $(14)^0 =$
7. $(1)^7 =$
8. $\dfrac{10^0 \times 1}{1} =$

5-11 Combined Operations

The following problems combine multiplication, division, and addition for powers of 10. Do one operation at a time and use the result for the next operation. It usually is helpful to determine the powers of 10 separately from the calculations for the coefficients.

Note that parentheses can be used instead of the multiplication sign. Numbers not separated by a + or − sign are factors to be multiplied.

Practice Problems 5-T
Answers at End of Chapter

Answers should be in scientific or engineering notation.

1. $(2 \times 10^6)(3 \times 10^3) \div (4 \times 10^4) =$
2. $(2 \times 10^6)(3 \times 10^{-3}) \div (4 \times 10^4) =$
3. $(2 \times 10^6)(3 \times 10^3)(4 \times 10^{-4}) =$
4. $(\sqrt{10^8})(10^3) =$
5. $\sqrt{10^4 \times 10^6} =$
6. $(25\sqrt{10^8}) + (3 \times 10^4) =$
7. $\dfrac{10^7}{\sqrt{10^7 \times 10^3}} =$
8. $\dfrac{1}{\sqrt{10^{-3} \times 10^{-5}}} =$
9. $(3 \times 10^3)(2 \times 10^2) + (3 \times 10^5) =$
10. $\dfrac{(4 \times 10^3)(2 \times 10^4)(3 \times 10^5)}{(8 \times 10^8)(1 \times 10^4)} =$

Review Problems
Answers to Odd-Numbered Problems at Back of Book

The following problems summarize operations with powers of 10.

1. $(3 \times 10^3) \times (2 \times 10^2) =$
2. $(3 \times 10^3) \div (2 \times 10^2) =$
3. $(5 \times 10^5) + (3 \times 10^5) =$
4. $(5 \times 10^5) - (3 \times 10^5) =$
5. $(5 \times 10^5) + (3 \times 10^6) =$
6. $1/(5 \times 10^5) =$
7. $1/(0.2 \times 10^{-5}) =$
8. $(3 \times 10^4)^2 =$
9. $\sqrt{9 \times 10^8} =$
10. $2 \times \sqrt{9 \times 10^4} =$
11. $[2.92 \times (4.3)^5] \times 0 =$
12. $5 \times 10^0 =$
13. $(18 \times 10^8) \div (2 \times 10^2) =$
14. $(3 \times 10^3)^2 =$

15. $\sqrt{9 \times 10^6} =$

16. $\sqrt[3]{8 \times 10^{18}} =$

17. $(3 \times 10^3)^2 + 4.2 \times 10^6 =$

18. $\sqrt{16 \times 10^6} + \sqrt[3]{8 \times 10^9} =$

19. $1/(5 \times 10^{-3}) + (4 \times 10^3) =$

20. $(4 \times 10^4) \times (2 \times 10^{-4}) =$

Answers to Practice Problems

5-A
1. 10^2
2. 10^3
3. 10^4
4. 10^5
5. 10^6
6. 10^7

5-B
1. 1000
2. 100
3. 1,000,000
4. 10,000

5-C
1. 10^{-3}
2. 10^{-3}
3. 10^{-2}
4. 10^2
5. 10^{-6}
6. 10^{-6}
7. 10^{-2}
8. 10^{-4}

5-D
1. 0.001
2. 0.1
3. 0.01
4. 0.000 001

5-E
1. $1/1000$
2. $1/10$
3. $1/100$
4. $1/1,000,000$

5-F
1. 10^4
2. 10^{-2}
3. 10^{-5}
4. 10^7
5. 10^8
6. 10^{-6}
7. 10^{-4}
8. 10^6

5-G
1. 10^2
2. 10^{-4}
3. 10^{-7}
4. 10^5
5. 10^6
6. 10^{-8}
7. 10^{-6}
8. 10^4

5-H See Problems 5-I for scientific notation. For answers in engineering notation:
1. 0.4×10^3
2. 0.47×10^3
3. 4×10^3
4. 4.7×10^3
5. 8×10^6
6. 40×10^{-3}
7. 47×10^{-3}
8. 4×10^{-3}
9. 4.7×10^{-3}
10. 8×10^{-6}

5-I See Problems 5-H

5-J
1. 6×10^6
2. 8×10^2
3. 7×10^8
4. 6×10^{15}
5. 5×10^7
6. 6×10^5
7. 7×10^4
8. 5×10^8
9. 6×10^6
10. 6×10^5

5-K
1. 6×10^{-6}
2. 8×10^{-2}
3. 7×10^{-8}
4. 6×10^{-15}
5. 5×10^{-7}
6. 6×10^{-5}
7. 7×10^{-4}
8. 5×10^{-8}
9. 9×10^{-7}
10. 1×10^{-2}

5-L
1. 8×10^6
2. 15×10^6
3. 6×10^{-3}
4. 28×10^{-3}
5. 8×10^2
6. 6×10^{-2}
7. 12×10^6

8. 8×10^{12}
9. 3×10^{-6}
10. 8
11. 14×10^6
12. 12×10^{-6}

5-M
1. 4×10^6
2. 3×10^6
3. 3.5×10^3
4. 1.5×10^2
5. 2.5×10^3
6. 4×10^6
7. 7×10^2
8. 2×10^4
9. 2×10^2
10. 1×10^2

5-N
1. 1×10^2
2. 1×10^{-2}
3. 1×10^{14}
4. 1×10^{-14}
5. 1×10^2
6. 3×10^2
7. 3×10^7
8. 2×10^{-9}
9. 0.5×10^{-2}
10. 1×10^{-2}

5-O
1. 10^4
2. 10^{-4}
3. 2×10^2
4. 4×10^{-5}
5. 10^{-2}
6. 0.5×10^{-2}
7. 1×10^{-7}
8. 8×10^8

5-P
1. 8×10^3
2. 4×10^3
3. 8×10^{-3}
4. 8×10^{-3}
5. 2.6×10^4
6. 2.6×10^4
7. 4.5×10^{-2}
8. 4.5×10^{-2}
9. 3×10^{-3}

10. 3×10^{-3}

5-Q
1. 1×10^6
2. 1×10^6
3. 1×10^{12}
4. 1×10^{14}
5. 10^{-6}
6. 49×10^6
7. 1×10^{14}
8. 16×10^{-12}
9. 4×10^4
10. 8×10^6

5-R
1. 2×10^4
2. 3×10^3
3. 10^2
4. 1×10^5
5. 1×10^3
6. 3×10^{-3}
7. 1×10^4
8. 6×10^3
9. 2×10^3
10. 5×10^4

5-S
1. 0
2. 10
3. 5
4. 1
5. 0
6. 1
7. 1
8. 1

5-T
1. 1.5×10^5, or 0.15×10^6
2. 1.5×10^{-1}, or 150×10^{-3}
3. 24×10^5, or 2.4×10^6
4. 10^7, or 10×10^6
5. 10^5, or 0.1×10^6
6. 28×10^4, or 280×10^3
7. 10^2, or 0.1×10^3
8. 10^4, or 10×10^3
9. 9×10^5, or 0.9×10^6
10. 3

6 LOGARITHMS

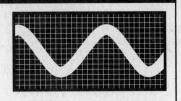

Logarithms are exponents. For example, in $10^2 = 100$ the exponent 2 is the logarithm of the number 100 with 10 as the base. Similarly, any number has a logarithm that is the required power of 10. Logarithms to base 10 are called *common logarithms* or Briggs' logarithms, named after the man who developed the first table of logarithms in the year 1624.

Another system of values is called *natural logarithms,* with the base 2.7183. In electricity and electronics this number is given the letter symbol *e*. In mathematics textbooks the Greek letter ϵ (epsilon) is sometimes used to represent this important constant.

Common logarithms to base 10 are indicated as $\log_{10}$ or just log. The natural logs are abbreviated ln. Common logs are generally used for numerical calculations. Their advantage is that they compress a large range of values to a much smaller range. For instance, all numbers between 10 and 1,000,000 have $\log_{10}$ values between 1 and 6. Natural logarithms are useful in formulas that describe many physical and chemical processes in nature. The values for one type of logarithm, however, can be converted to the other.

The need for log tables in numerical calculations has practically disappeared now with the use of the electronic calculator. A scientific calculator can quickly give the logarithm, either as $\log_{10}$ or ln, for any number. Just use the $\boxed{\log}$ key for the common logarithm, and the $\boxed{\ln}$ or $\boxed{\ln_x}$ key for the natural logarithm.

However, it is still necessary to appreciate the meaning of logarithms because they are used in many applications of electronics. One important example is the decibel (dB), which is a logarithmic unit commonly used to compare two values of electric power or indicate relative signal strengths. The use of logarithmic units has the advantage of compressing a wide range of values.

6-1   Examples of Logarithms

The following values illustrate how $\log_{10}$ values are the same as powers of 10:

$$10^1 = 10 \text{ or } 1 \text{ is log } 10$$
$$10^2 = 100 \text{ or } 2 \text{ is log } 100$$
$$10^3 = 1000 \text{ or } 3 \text{ is log } 1000$$
$$10^4 = 10,000 \text{ or } 4 \text{ is log } 10,000$$
$$10^5 = 100,000 \text{ or } 5 \text{ is log } 100,000$$
$$10^6 = 1,000,000 \text{ or } 6 \text{ is log } 1,000,000$$

Where no subscript is indicated for log, it is to base 10. In general, the formula for logarithms is

$$N = B^l \tag{6-1}$$

where the exponent l is the logarithm of the number N with base B. For instance, for N of 1000, l is 3 with B of 10, or the log of 1000 is 3.

The same system applies to decimal fractions, which are negative powers of 10. As examples:

$$0.1 = 10^{-1} \text{ or } -1 \text{ is log } 0.1$$
$$0.01 = 10^{-2} \text{ or } -2 \text{ is log } 0.01$$
$$0.001 = 10^{-3} \text{ or } -3 \text{ is log } 0.001$$
$$0.0001 = 10^{-4} \text{ or } -4 \text{ is log } 0.0001$$
$$0.000\,01 = 10^{-5} \text{ or } -5 \text{ is log } 0.000\,01$$
$$0.000\,001 = 10^{-6} \text{ or } -6 \text{ is log } 0.000\,001$$

Note that decimal fractions less than 1 have negative logarithms because the fractions have negative exponents expressed as a power of 10.

Practice Problems 6-A
Answers at End of Chapter

Give $\log_{10}$ for the following numerical values.

1.	100 *2*	**5.**	10,000 *4*
2.	0.01 *-2*	**6.**	100,000 *5*
3.	1000 *3*	**7.**	1,000,000 *6*
4.	0.001 *-3*	**8.**	10,000,000 *7*

6-2 Characteristic and Mantissa of a Logarithm

All the examples so far have logarithms that are whole numbers, because these exponents are exact powers of 10. However, any number can be given a value of $\log_{10}$ because any number can be described as a power of 10. The exponent can be a fractional value like 1.8, instead of a whole number. Still, numbers greater than 1 always have a positive log, while decimal fractions less than 1 have a negative log.

The number 63, as an example, has the log of 1.8, approximately. This can be stated as

$$10^{1.8} = 63 \text{ (approx)}$$

In other words 10 raised to the 1.8 power is approximately equal to 63.

Consider the progression of log values for numbers. The log of 10 is 1 and the log of 100 is 2. Since 63 is between 10 and 100, the log of 63 is between 1 and 2. The log gets closer to 2 as the number gets closer to 100. However, the progression is not uniform. You must look up the exponent in a log table or use an electronic calculator.

If there is a question of how the value of a fractional exponent can be calculated, the answer is that any number can be raised to any power. The calculated value can be obtained graphically or algebraically from a geometric progression of terms equivalent to the exponential function.

A log value such as 1.8 must be considered in two parts. The whole number part, 1 here, is separate because it tells how many decimal places are in the number. It shows that the number 63 is between 10 and 100. This part is called the *characteristic* of the logarithm. The fractional part, 0.8, is called the *mantissa*. It depends on the digits in the number. For the log value 1.8, the characteristic is 1 and the mantissa is 0.8.

To emphasize the difference between the characteristic and the mantissa, consider these values. The numbers 20, 30, 40, 50, 60, 70, 80 and 90 all have the same characteristic, 1, because they all have two decimal places. All these numbers are more than 10 but less than 100. In general, to find the log of numbers greater than 1, the characteristic is one less than the number of decimal places. The mantissas for these numbers, however, increase as the digits increase from 2 to 9. Specifically, these log values with increasing mantissas are:

log 20 = 1.301	log 50 = 1.699	log 80 = 1.903
log 30 = 1.477	log 60 = 1.778	log 90 = 1.954
log 40 = 1.602	log 70 = 1.845	

The numbers 63, 630, 6300, and 63,000 all have the same mantissa because each has the digits 6 and 3. However, the characteristic increases from 1 to 4 as the number of decimal places increases.

In summary, any positive number greater than 1 has a positive logarithm, in two parts. For the example of log 63 = 1.8, then,

1. The characteristic is to the left of the decimal point in the logarithm, to indicate how many decimal places are in the number. Subtract 1 from the number of places in the number for the characteristic. For log 63 = 1.8, the 1 is the characteristic.
2. The mantissa is to the right of the decimal point in the logarithm. The value of the mantissa is found from log tables or with an electronic calculator. For log 63 = 1.8, the 0.8 is the mantissa.

Finding the Characteristic A good way to visualize the characteristic for the log of a number is to write the number in scientific notation, as a power of 10 with a coefficient more than 1 but less than 10. Then the coefficient always has one decimal place to the left of the decimal point. In this form the power of 10 is the characteristic in the logarithm. See Table 6-1.

Table 6-1 Logarithms for Numbers

Number	Scientific Notation	Characteristic	Mantissa*	Complete Logarithm
6.3	6.3×10^0	0	0.8	0.8
63	6.3×10^1	1	0.8	1.8
630	6.3×10^2	2	0.8	2.8
6300	6.3×10^3	3	0.8	3.8

*Mantissa of 0.8 is approximate value from log table or electronic calculator.

Practice Problems 6-B
Answers at End of Chapter

Give only the characteristic for the log of the following numbers.

1.	3500	**5.**	4.625×10^4	
2.	3.5×10^3	**6.**	46,250	
3.	0.002	**7.**	8.3×10^9	
4.	2×10^{-3}	**8.**	8.3×10^{-9}	

Finding the Complete Logarithm When the characteristic is known, the mantissa can be found from a log table, like the one shown on pages 54 and 55. Look for the first two digits of the number in the vertical column at the left and the third digit in the horizontal row at the top. For instance, the mantissa for the digits 222 is 0.3464. The mantissa is always a fraction after the decimal point in the log. Whether the digit in front of the decimal point is 0 or more depends on the characteristic for the number of decimal places in the number whose log we want to find. For instance,

log 2.22 is 0.3464
log 22.2 is 1.3464
log 222 is 2.3464

When you use an electronic calculator to find the logarithm, the displayed value includes both the characteristic and the mantissa. This method is certainly much easier. Just punch in the number on the keyboard and press [log].

On some calculators the $\log_{10}$ and ln functions are on the same key, but you must press the "function" key first to find the natural log of the number.

Practice Problems 6-C
Answers at End of Chapter

Give the complete logarithm to three decimal places, with characteristic and mantissa, for the following numbers.

1.	2	**7.**	8320
2.	5	**8.**	9000
3.	8	**9.**	14,000
4.	40	**10.**	33,000
5.	620	**11.**	84,200
6.	6400	**12.**	164,700

6-3 Negative Logarithms

As discussed before, decimal fractions less than 1 have negative logarithms. For example, the log of 0.01 or 10^{-2} is -2. A problem arises, though, when the negative log has a mantissa that is not zero. The reason is that a mantissa is always positive. Actually, the mantissa in the logarithm of a number only measures the progression of the digits from 1111 to 9999, without any relation to the decimal place. Consider the example of log 200 for a positive log. Since $200 = 2 \times 10^2$, the characteristic is 2 and the mantissa is 0.301 for the complete logarithm 2.301. As a comparison, consider the example of log 0.02 or 2×10^{-2}. The characteristic now is -2, but the mantissa is still 0.301. We can write this log as

$$\log 0.02 = \bar{2}.301$$

The bar over the 2 shows that the characteristic is negative, or -2, but the mantissa of 0.301 is still positive. Although the number is written as a single unit, remember that it actually represents a negative characteristic (-2) added to a positive mantissa (0.301) or

$$-2. + 0.301$$

This method of showing a negative characteristic can be inconvenient when working with logarithms, so it is common practice to rewrite the logarithm by adding 10 and subtracting 10 as follows:

$$(\bar{2}.301 + 10) - 10$$

The value is not changed, since we are adding and subtracting the same number. However, since $-2 + 10 = 8$, the log can now be written as

$$8.301 - 10$$

The number 8.301 now represents a single unit which is positive, to be combined with -10. Then

$$\begin{array}{r} -10.000 \\ +8.301 \\ \hline -1.699 \end{array}$$

In the form -1.699, the entire value is negative for log 0.02. This is how the electronic calculator displays the log for decimal fractions.

It may be of interest to see why the log shifts from a characteristic of −2 for the log of 0.01 to a characteristic of −1 for the log of 0.02. After all, 0.02 is a larger number than 0.01. Remember that 0.02 is $^2/_{100}$ and 0.01 is $^1/_{100}$. The larger number must have a smaller negative log because we are dealing with negative exponents. The larger the negative exponent of a number, the smaller the number is. As further examples, 10^{-3} or one thousandth is more than 10^{-6} or one millionth.

Practice Problems 6-D
Answers at End of Chapter

Give the negative logarithm to three decimal places as displayed on an electronic calculator for the following numbers.

−1.699

1.	0.02	**5.**	0.005
2.	0.05	**6.**	0.009
3.	0.09	**7.**	0.000 066
4.	0.003	**8.**	0.000 006 8

6-4 Antilogarithms

An antilog is the number N represented by the log l in the formula $N = B^l$. As an example, for base 10 and $N = 100$ or 10^2, the log is 2 and the antilog is 100. In order to find the antilog, reverse the process of finding the log. The two steps for finding the antilog from a log table are

1. Use just the mantissa of the log to find the digits of the number N represented by that mantissa. For instance, if the mantissa is 0.301, the digits in the number are 200, without regard to the decimal point.
2. In order to have the correct number of decimal places, write the digits from the mantissa in scientific notation with a power of 10 equal to the characteristic in the logarithm. As examples,

Logarithm	Antilogarithm
1.301	$2.0 \times 10^1 = 20$
2.301	$2.0 \times 10^2 = 200$
3.301	$2.0 \times 10^3 = 2000$
4.301	$2.0 \times 10^4 = 20,000$

This system applies to positive logarithms or to a negative characteristic with a positive mantissa, but not to negative logarithms.

An electronic calculator makes it easy to find the antilog. Just punch in the logarithm as a numerical value on the keyboard and push the antilog key to display the corresponding number. The antilog key for common logarithms is generally labeled $\boxed{10^x}$.

For the antilog of 2.301, as an example, punch in these digits, including characteristic and mantissa, with the required decimal point. Then push $\boxed{10^x}$. The answer displayed will be 199.99 or 200, equal to 2×10^2.

With negative logarithms, as used for decimal fractions, use the ± invert key *after* punching in the log value. Be sure you see the negative sign displayed for the log value before you press the antilog key. Then −1.699 for the log yields 0.0199 or approximately 0.02 as the answer.

You cannot use a negative characteristic with a positive mantissa on the calculator. For instance $\overline{2}.301$ must be converted to the form −1.699.

Practice Problems 6-E
Answers at End of Chapter

Find the antilog for the following log values:

1000 *0.001* *7.998 or 8*

1.	3.0	**6.**	0.3464
2.	−3.0	**7.**	1.3464
3.	2.301	**8.**	5.3464
4.	−1.699	**9.**	0.602
5.	$\overline{2}.301$	**10.**	0.903

6-5 Logarithmic Graph Paper

For many applications in electronics, a curve or graph is shown on a logarithmic scale, instead of using uniform spacing of the grid lines in the graph paper. Examples are shown in Figs. 6-1 and 6-2 for a logarithmic scale on the horizontal x axis. In this system, the results are plotted on the graph for values that are powers of 10, such as tens, hundreds, and thousands. The advantage is that a much wider range of values can be shown on the graph. To compensate for the fact that the scale of values increases exponentially, the spacing of the grid lines is made proportional to the logarithm of the scale values. Actually, only the mantissas from 1 to 10 are used for the scaling, since multiple numbers repeat the same mantissas with a different characteristic.

An expanded view of logarithmic spacing of the grid lines for one cycle of numbers from 1 to 10 is shown in Fig. 6-1. The framework of all the lines is called the *grid*. Note that the x axis is divided into ten parts, with

Table of Logarithms (Four-Place Mantissas)

No.	0	1	2	3	4	5	6	7	8	9
10	0000	0043	0086	0128	0170	0212	0253	0294	0334	0374
11	0414	0453	0492	0531	0569	0607	0645	0682	0719	0755
12	0792	0828	0864	0899	0934	0969	1004	1038	1072	1106
13	1139	1173	1206	1239	1271	1303	1335	1367	1399	1430
14	1461	1492	1523	1553	1584	1614	1644	1673	1703	1732
15	1761	1790	1818	1847	1875	1903	1931	1959	1987	2014
16	2041	2068	2095	2122	2148	2175	2201	2227	2253	2279
17	2304	2330	2355	2380	2405	2430	2455	2480	2504	2529
18	2553	2577	2601	2625	2648	2672	2695	2718	2742	2765
19	2788	2810	2833	2856	2878	2900	2923	2945	2967	2989
20	3010	3032	3054	3075	3096	3118	3139	3160	3181	3201
21	3222	3243	3263	3284	3304	3324	3345	3365	3385	3404
22	3424	3444	3464	3483	3502	3522	3541	3560	3579	3598
23	3617	3636	3655	3674	3692	3711	3729	3747	3766	3784
24	3802	3820	3838	3856	3874	3892	3909	3927	3945	3962
25	3979	3997	4014	4031	4048	4065	4082	4099	4116	4133
26	4150	4166	4183	4200	4216	4232	4249	4265	4281	4298
27	4314	4330	4346	4362	4378	4393	4409	4425	4440	4456
28	4472	4487	4502	4518	4533	4548	4564	4579	4594	4609
29	4624	4639	4654	4669	4683	4698	4713	4728	4742	4757
30	4771	4786	4800	4814	4829	4843	4857	4871	4886	4900
31	4914	4928	4942	4955	4969	4983	4997	5011	5024	5038
32	5051	5065	5079	5092	5105	5119	5132	5145	5159	5172
33	5185	5198	5211	5224	5237	5250	5263	5276	5289	5302
34	5315	5328	5340	5353	5366	5378	5391	5403	5416	5428
35	5441	5453	5465	5478	5490	5502	5514	5527	5539	5551
36	5563	5575	5587	5599	5611	5623	5635	5647	5658	5670
37	5682	5694	5705	5717	5729	5740	5752	5763	5775	5786
38	5798	5809	5821	5832	5843	5855	5866	5877	5888	5899
39	5911	5922	5933	5944	5955	5966	5977	5988	5999	6010
40	6021	6031	6042	6053	6064	6075	6085	6096	6107	6117
41	6128	6138	6149	6160	6170	6180	6191	6201	6212	6222
42	6232	6243	6253	6263	6274	6284	6294	6304	6314	6325
43	6335	6345	6355	6365	6375	6385	6395	6405	6415	6425
44	6435	6444	6454	6464	6474	6484	6493	6503	6513	6522
45	6532	6542	6551	6561	6571	6580	6590	6599	6609	6618
46	6628	6637	6646	6656	6665	6675	6684	6693	6702	6712
47	6721	6730	6739	6749	6758	6767	6776	6785	6794	6803
48	6812	6821	6830	6839	6848	6857	6866	6875	6884	6893
49	6902	6911	6920	6928	6937	6946	6955	6964	6972	6981
50	6990	6998	7007	7016	7024	7033	7042	7050	7059	7067
51	7076	7084	7093	7101	7110	7118	7126	7135	7143	7152
52	7160	7168	7177	7185	7193	7202	7210	7218	7226	7235
53	7243	7251	7259	7267	7275	7284	7292	7300	7308	7316
54	7324	7332	7340	7348	7356	7364	7372	7380	7388	7396
No.	0	1	2	3	4	5	6	7	8	9

Table of Logarithms (continued)

No.	0	1	2	3	4	5	6	7	8	9
55	7404	7412	7419	7427	7435	7443	7451	7459	7466	7474
56	7482	7490	7497	7505	7513	7520	7528	7536	7543	7551
57	7559	7566	7574	7582	7589	7597	7604	7612	7619	7627
58	7634	7642	7649	7657	7664	7672	7679	7686	7694	7701
59	7709	7716	7723	7731	7738	7745	7752	7760	7767	7774
60	7782	7789	7796	7803	7810	7818	7825	7832	7839	7846
61	7853	7860	7868	7875	7882	7889	7896	7903	7910	7917
62	7924	7931	7938	7945	7952	7959	7966	7973	7980	7987
63	7993	8000	8007	8014	8021	8028	8035	8041	8048	8055
64	8062	8069	8075	8082	8089	8096	8102	8109	8116	8122
65	8129	8136	8142	8149	8156	8162	8169	8176	8182	8189
66	8195	8202	8209	8215	8222	8228	8235	8241	8248	8254
67	8261	8267	8274	8280	8287	8293	8299	8306	8312	8319
68	8325	8331	8338	8344	8351	8357	8363	8370	8376	8382
69	8388	8395	8401	8407	8414	8420	8426	8432	8439	8445
70	8451	8457	8463	8470	8476	8482	8488	8494	8500	8506
71	8513	8519	8525	8531	8537	8543	8549	8555	8561	8567
72	8573	8579	8585	8591	8597	8603	8609	8615	8621	8627
73	8633	8639	8645	8651	8657	8663	8669	8675	8681	8686
74	8692	8698	8704	8710	8716	8722	8727	8733	8739	8745
75	8751	8756	8762	8768	8774	8779	8785	8791	8797	8802
76	8808	8814	8820	8825	8831	8837	8842	8848	8854	8859
77	8865	8871	8876	8882	8887	8893	8899	8904	8910	8915
78	8921	8927	8932	8938	8943	8949	8954	8960	8965	8971
79	8976	8982	8987	8993	8998	9004	9009	9015	9020	9025
80	9031	9036	9042	9047	9053	9058	9063	9069	9074	9079
81	9085	9090	9096	9101	9106	9112	9117	9122	9128	9133
82	9138	9143	9149	9154	9159	9165	9170	9175	9180	9186
83	9191	9196	9201	9206	9212	9217	9222	9227	9232	9238
84	9243	9248	9253	9258	9263	9269	9274	9279	9284	9289
85	9294	9299	9304	9309	9315	9320	9325	9330	9335	9340
86	9345	9350	9355	9360	9365	9370	9375	9380	9385	9390
87	9395	9400	9405	9410	9415	9420	9425	9430	9435	9440
88	9445	9450	9455	9460	9465	9469	9474	9479	9484	9489
89	9494	9499	9504	9509	9513	9518	9523	9528	9533	9538
90	9542	9547	9552	9557	9562	9566	9571	9576	9581	9586
91	9590	9595	9600	9605	9609	9614	9619	9624	9628	9633
92	9638	9643	9647	9652	9657	9661	9666	9671	9675	9680
93	9685	9689	9694	9699	9703	9708	9713	9717	9722	9727
94	9731	9736	9741	9745	9750	9754	9759	9763	9768	9773
95	9777	9782	9786	9791	9795	9800	9805	9809	9814	9818
96	9823	9827	9832	9836	9841	9845	9850	9854	9859	9863
97	9868	9872	9877	9881	9886	9890	9894	9899	9903	9908
98	9912	9917	9921	9926	9930	9934	9939	9943	9948	9952
99	9956	9961	9965	9969	9974	9978	9983	9987	9991	9996
No.	0	1	2	3	4	5	6	7	8	9

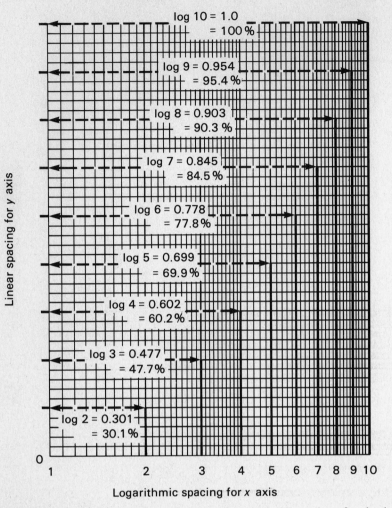

Fig. 6-1 Graph with logarithmic spacing for *x* axis. The values for the logs indicate spacing from the origin relative to log 10.

the grid lines relatively far apart at the left near 1 and 2 and more crowded for higher values toward 9. The reason is that the horizontal distance or spacing from the start at the origin is made proportional to the logarithm of the *x* value. For instance, log 2 is 0.301; thus the grid line for *x* = 2 is approximately three-tenths of the entire distance across to 10. Similarly, the grid line for *x* = 3 is almost one-half the distance across, as log 3 = 0.477. The same idea applies to all the other grid lines for *x* values from 1 to 10. This spacing would also apply to multiples, such as 10 to 100 or 100 to 1000. Note that with log spacing, the *x* values start from 1 rather than 0. With logarithmic values, the 1 corresponds to 0, as $10^0 = 1$ or 0 is the log of 1.

The group of values illustrated on the *x* axis in Fig. 6-1 is called a *cycle* of log spacing. In typical log paper

there are usually three or four cycles or repeats of the log spacing, for multiples of tens, hundreds, thousands, etc., for plotting a wide range of values. The graph in Fig. 6-2 is plotted on four-cycle *semilogarithmic* (semilog) paper. The four cycles of values are tens, hundreds, thousands, and ten thousands. Semilog means that only one axis, either *x* or *y*, has log spacing, while the other axis has the usual linear or uniform spacing. When both axes have logarithmic spacing, the result is *log-log* or full logarithmic graph paper.

Semilog paper is used for a graph when one set of values has a very wide range of values but the other does not. An example is the audio-frequency response curve shown in Fig. 6-2. The vertical *y* axis with linear spacing represents signal voltage output from an audio amplifier, in volts. The output varies from 1 to 8 V,

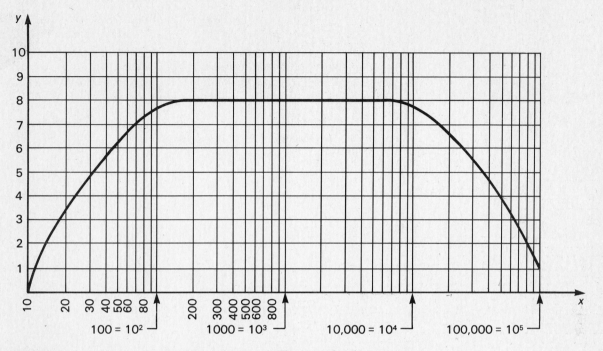

Fig. 6-2 Curve plotted on four-cycle semilog paper. The x axis has log spacing. The heavy line is known as an audio-frequency response curve.

which is a range that can easily be shown with linear spacing. However, the horizontal *x* axis represents audio frequencies in units of hertz (Hz). This range is 10 to 100,000 Hz. Note that the first cycle of log values is tens, then hundreds, thousands, and ten thousands. As an example, the next-to-last *x* value near the end of the graph at the right represents the frequency of 90,000 Hz.

The advantage of this logarithmic scaling for the frequency values on the *x* axis is the tremendous range that can be shown in a compact form. If this range, from 10 to 100,000 Hz, were plotted on a linear scale in units of tens, the graph would be much too wide. At the opposite extreme, if the graph were plotted in steps of thousands of hertz, much of the information for frequencies below 1000 Hz would be lost. As a result, response curves for audio-frequency equipment are usually shown on semilog graph paper.

Practice Problems 6-F
Answers at End of Chapter

For the graph in Fig. 6-2, give the *y* value for the following values of *x*.

1. 10
2. 20
3. 45
4. 200

5. 1000
6. 2000
7. 8000
8. 10,000
9. 20,000
10. 50,000

6-6 Decibel (dB) Units

The decibel unit is commonly used to compare two levels of electric power in terms of the logarithm of their ratio. The formula is

$$dB = 10 \times \log\left(\frac{P_2}{P_1}\right) \qquad (6\text{-}2)$$

In this formula P_2 and P_1 are the two values of power. For instance, let P_2 be 2 watts (W) and P_1 1 W. Then

$$dB = 10 \log\left(\frac{2\ W}{1\ W}\right) = 10 \times \log(2)$$
$$= 10 \times 0.301$$
$$dB = 3 \text{ (approx)}$$

The advantage is that the decibel units compress a wide range of values being compared. As an example, the range of 1000 to 1 in power is equal to 30 dB.

In the decibel formula, any unit of power can be used, but P_2 and P_1 must be in the same unit, so that they can cancel. Remember that 1 milliwatt (mW) is 1×10^{-3} W.

Also, let P_2 be the larger value in the ratio. Then the log is always positive. This way, there is no need to work with negative logarithms. An increase of power is considered as +dB, dB up, or a dB gain. The + sign is usually omitted. A decrease of power is indicated as −dB, dB down, or a −dB loss after the value has been found from a positive logarithm.

Example   An audio amplifier has input power of 200 mW and output of 400 mW. Calculate the decibel gain.

Answer $dB = 10 \times \log\left(\dfrac{400 \text{ mW}}{200 \text{ mW}}\right)$

$= 10 \times \log 2$

$= 10 \times 0.301$

$dB = 3$

This answer means a gain of 3 dB. In general, any case of doubling the power means a 3-dB gain.

Example   An audio line has input power of 14 mW and output of 7 mW. Calculate the decibel loss.

Answer $dB = 10 \log\left(\dfrac{14 \text{ mW}}{7 \text{ mW}}\right) = 10 \times \log 2$

$= 10 \times 3.01$

$dB = 3$

In this case, the answer is −3 dB for the loss of power. In general, any case of one-half the power means a loss of 3 dB.

The general procedure for calculating decibel values is as follows:

1. To calculate the ratio of the two levels of power, divide P_2 by P_1. Use the same units for both values. Make the larger power value P_2.
2. Find the logarithm of the numerical ratio.
3. Multiply the logarithm by 10.

Multiplying the logarithm of the power ratio by 10 gives the answer in decibels. The reason for multiplying by 10 is that a decibel is one-tenth of a bel. Do not use any antilogarithms, since the decibel is a logarithmic unit. A gain in power means the answer is in +dB units; a loss in power corresponds to −dB units.

Practice Problems 6-G
Answers at End of Chapter

Find the gain or loss in decibels for the following power ratios.

1. 4 W input and 8 W output
2. 8 W input and 16 W output
3. 4 W input and 16 W output
4. 4 W input and 40 W output
5. 300 mW input and 0.6 W output
6. 36 W input and 18 W output
7. 36 W input and 9 W output
8. 17 mW input and 1700 mW output
9. 1.43 W input and 9.65 W output
10. 72 mW input and 20 W output
11. 2 mW input and 2 W output
12. 3 W input and 1.47 mW output

6-7   Natural Logarithms

Remember that logarithms are just a system of exponents for a specified base. The base 10 is commonly used for logarithms in numerical calculations because of our decimal number system. In this system, the values are called *common logarithms*. However, another system of logarithms often used in scientific work has a base of 2.7183. This special number is the value for the limit of the natural function

$$\left(1 + \frac{1}{x}\right)^x$$

as x increases toward infinity (Fig. 6-3). The letter e is used as the symbol for 2.7183. Note that the curve in Fig. 6-3 rises to this value. (Some mathematics books use the Greek letter epsilon ϵ as the symbol for the base of the natural logarithms.) Logarithms to this base are called *natural logs*. They are indicated as ln for natural log. No subscript means the log is to base 10. The system of natural logs was invented by John Napier (1550–1617). Sometimes they are called napierian logarithms.

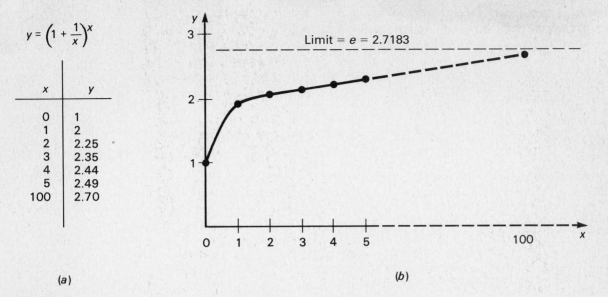

$y = \left(1 + \dfrac{1}{x}\right)^{x}$

x	y
0	1
1	2
2	2.25
3	2.35
4	2.44
5	2.49
100	2.70

(a)

(b)

Fig. 6-3 The natural function $y = (1 + 1/x)^{x}$ in tabular form and as a linear graph. (a) Table of values for x and y. (b) Graph of curve approaching $e = 2.7183$ as a limit.

Consider the table of values and corresponding graph in Fig. 6-3 for the function

$$y = \left(1 + \frac{1}{x}\right)^{x}$$

When x is 1, y is $1 + 1 = 2$ to the first power. Then $y = 2$.

For x of 2, y is $1 + \frac{1}{2} = 1.5$ to the second power. This y value is $(1.5)^{2} = 2.25$.

Also, when x is 3, y is $1 + \frac{1}{3} = 1.33$ to the third power. Then y is $(1.33)^{3} = 2.35$. Notice that this value is not much more than 2.25, the x value of 2. As x increases toward the highest possible number, the y value only approaches the value of 2.7183. Therefore, $y = 2.7183$ is the limit for the value of the function.

The equation

$$y = \left(1 + \frac{1}{x}\right)^{x}$$

is a natural function because it describes the way many processes occur in nature. Physically, the statement says that increases occur at a slower rate because of the previous increases. An example is charging a capacitor. The more charge that is put into the capacitor, the slower is the rate at which it takes on more charge.

To illustrate how we can have logarithms to base e, consider the following values, with e rounded off to 2.7 to determine ln:

Number	Log$_{10}$	Ln
$2.7 = (2.7)^{1}$	0.431	1
$10 = (10)^{1}$	1	2.3
$7.29 = (2.7)^{2}$	0.86	2 (approx)
$100 = (10)^{2}$	2	4.6

These logarithms have a positive characteristic and mantissa. For numbers with a negative logarithm,

Number	Log$_{10}$	Ln
$0.370 = 1/2.7$	−0.43	−1.0 (approx)
$0.1 = 1/10$	−1.0	−2.3
$0.137 = 1/(2.7)^{2}$	−0.86	−2.0 (approx)
$0.01 = 1/100$	−2.0	−4.6

There are tables of natural logs, but most scientific type electronic calculators have provision for finding these logs and their antilogs. To find the natural log, punch in the number on the keyboard and then press $\boxed{\text{ln}}$. The natural log of the number is then displayed.

Example   Find the natural log of 98.7.

Answer   Using $\boxed{\text{ln}}$, the answer is 4.592.

To find the antilog, the required key is generally labeled e^{x} on the electronic calculator. Enter the value of the log on the keyboard and press $\boxed{e^{x}}$.

It should be noted that some calculators require pressing the second function [2ndF] key for natural logs while other calculators use [2ndF] for common logs.

> **Example** Find the antilog for the natural log 4.592.
>
> **Answer** Using e^x, the answer is 98.7.

For a negative log, press [±] to change the sign of the log value before pressing the antilog key.

> **Example** Find the antilog for the natural log −4.592.
>
> **Answer** Using e^x, the answer is 0.01.

It is possible to convert log values between the two systems, since ln 10 is 2.3 and $\log_{10}$ 2.7183 is 0.43. The formulas are

$$\ln x = 2.3 \log_{10} x$$
$$\log_{10} x = 0.43 \ln x$$

For example:

$$\ln 1000 = 6.9 \text{ or } 2.3 \times 3$$
$$\log 1000 = 3 \text{ or } 6.9 \times 0.43$$

The 3 for $\log_{10}$ and 6.9 for ln are for the same number, 1000. Some of these calculations are only approximate because more decimal places must be used to get the exact value. Actually, the conversion factor of 0.43 is equal to $\frac{1}{2.3}$.

Remember that ln x will always be larger than $\log_{10} x$. The reason is that e is a smaller base, so it needs a larger exponent than base 10. Therefore, in converting to ln, multiply $\log_{10}$ by 2.3 to get a larger ln. For the opposite case, in converting from ln to $\log_{10}$, divide by 2.3 or multiply by 0.43 to get a smaller $\log_{10}$.

Practice Problems 6-H
Answers at End of Chapter

Give the value of the natural log (ln) for the following.

1. $N = 2.7183$ _1_
2. $N = 7.389$ _2_

3. $N = 20.086$ _3_
4. $N = 1000$ _7_
5. $N = 1,000,000$ _13.8_
6. $N = 468$ _6.15_
7. $N = 17,200$ _9.75_
8. $N = 7842$ _8.97_

Review Problems
Answers to Odd-Numbered Problems at Back of Book

Find $\log_{10}$ for the following numbers.

1. 0.2 _−0.699_
2. 2.0 _0.301_
3. 2.2 _0.342_
4. 2200 _3.342_
5. 6,400,000 _6.806_
6. 726,000 _5.861_
7. 48,000 _4.681_
8. 480 _2.681_
9. 37 _1.568_
10. 0.0037 _−2.432_

Find ln for the following numbers.

11. 8 _2.079_
12. 10 _2.302_
13. 2200 _7.696_
14. 480 _6.173_
15. 37 _3.611_

Find the antilog of the following common logarithms.

16. 0.699 _5_
17. 0.301 _1.999 or 2_
18. 0.342 _2.198_
19. 3.342 _2197.859_
20. 6.806 _6397448.355_ _6400000_

Find the gain or loss in + or − dB for the following power ratios.

21. 17 W input and 34 W output _3 dB_
22. 2.8 mW input and 5.6 mW output _3 dB_
23. 400 mW input and 100 mW output _−6.02 dB_
24. 9 mW input and 2 W output _23.468_
25. 8 mW input and 8 mW output _0_

Answers to Practice Problems

6-A
1. 2
2. −2
3. 3
4. −3
5. 4
6. 5
7. 6
8. 7

6-B
1. 3
2. 3
3. −3
4. −3
5. 4
6. 4
7. 9
8. −9

6-C
1. 0.301
2. 0.699
3. 0.903
4. 1.602
5. 2.792
6. 3.806
7. 3.920
8. 3.954
9. 4.146
10. 4.519

11. 4.925
12. 5.217

6-D
1. −1.699
2. −1.301
3. −1.046
4. −2.523
5. −2.301
6. −2.046
7. −4.180
8. −5.167

6-E
1. 1000
2. 0.001
3. 200 or 2×10^2
4. 0.02 or 2×10^{-2}
5. 0.02 or 2×10^{-2}
6. 2.22
7. 2.22×10 or 22.2
8. 222,000 or 2.22×10^5
9. 4.0 (approx)
10. 8.0 (approx)

6-F
1. 0
2. 3.3
3. 6.1
4. 8
5. 8
6. 8

7. 8
8. 7.8
9. 6.6
10. 4

6-G
1. 3 dB
2. 3 dB
3. 6 dB
4. 10 dB
5. 3 dB
6. −3 dB
7. −6 dB
8. 20 dB
9. 8.3 dB
10. 24.4 dB
11. 30 dB
12. −33 dB

6-H
1. 1.0
2. 2.0
3. 3.0
4. 6.9
5. 13.8
6. 6.148
7. 9.753
8. 8.967

7 THE METRIC SYSTEM

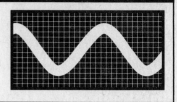

Metric units are generally used in scientific work because calculations are easier with decimal values based on multiples of 10. For example, 1 meter (m) is 100 centimeters (cm), or 1 m = 1 × 10² cm. Compare this with the U.S. customary system of units, in which 1 yard is 3 feet and 1 foot is 12 inches. Furthermore, the metric prefixes for powers of 10 are very common in electronics because typical values are much more or much less than the basic units. For instance, an electric current of 0.009 ampere (A) is equal to 9 milliamperes (mA), where milli (m) is the metric prefix for 10^{-3}. In general, the metric units with decimal multiples form an international system for measurements used throughout the world.

More details are explained in the following sections:

7-1 Metric Prefixes

These decimal multiples and submultiples are summarized in Table 7-1. They are in two groups. The group at the left is for positive powers of 10, such as mega for 10^6 and kilo for 10^3. At the right is the group for negative powers of 10, such as micro for 10^{-6} and milli for 10^{-3}. Each horizontal row includes the prefixes for the positive and negative exponent for the same power. For instance, mega (10^6) and micro (10^{-6}) are in the same horizontal row.

In the symbols, or abbreviations, note that lowercase letters are used for decimal fractions with negative exponents—all the metric prefixes in the group on the right-hand side of Table 7-1. An example is "m" for milli, equal to 0.001 or 10^{-3}. Also, μ (the Greek letter mu) is used for micro, equal to 10^{-6}. The capital (or uppercase) letters, such as M for mega or 1 million = 10^6, are used for multipliers greater than 1. However,

an exception is made for k for kilo = 10^3 because K is reserved for the kelvin unit of absolute temperature.

Examples of the use of the metric prefixes for units commonly used in electronics are given in Table 7-2. Abbreviations for the electrical units are:

A for ampere
V for volt
Ω for ohm
F for farad
H for henry
s for second
Hz for hertz

The examples in Table 7-2 indicate the wide range of values, which is why the metric prefixes are so important in electronics. A high value of voltage is 25 kV for the picture tube in a television receiver. At the opposite extreme, a low value is 25 μV for the radio signal picked up by a receiving antenna. With electric currents, the value can be high at 20 A, low at 8 mA, or even smaller at 7 μA. For resistance values, the R can be just a few ohms or many megohms (notice that the a in mega is not used). Finally, in frequencies the values can be the 60-Hz power-line frequency up to 2 kHz for an audio frequency or higher than 88 MHz for radio-frequency signals. The wide range of values expressed with metric prefixes is illustrated in Fig. 7-1.

In converting to a metric prefix, first convert the value to a coefficient with the required power of 10, then substitute the metric prefix.

Example   Convert 0.005 A to milliamperes.

Answer   0.005 A = 5 × 10^{-3} A = 5 mA

Example   Convert 7,000,000 Ω to megohms.

Answer   7,000,000 Ω = 7 × 10^6 Ω = 7 MΩ

Table 7-1 Metric Prefixes*

Positive Power of 10	Symbol	← Prefix	→ Prefix	Symbol	Negative Power of 10
10^{12}	T	tera	pico	p	10^{-12}
10^{9}	G	giga	nano	n	10^{-9}
10^{6}	M	mega	micro	μ	10^{-6}
10^{3}	k	kilo	milli	m	10^{-3}
10^{2}	h	hecto	centi	c	10^{-2}
10^{1}	da	deka	deci	d	10^{-1}

*The metric prefixes also include peta (P) for 10^{15}, exa (E) for 10^{18}, atto (a) for 10^{-18}, and femto (f) for 10^{-15}.

Practice Problems 7-A
Answers at End of Chapter

Convert the following to values with a metric prefix.

1. 0.009 A   _9 mA_
2. 5,000,000 Ω   _5 MΩ_
3. 0.000 006 s   _6 µs_
4. 4,000,000 Hz   _4 MHz_
5. 8000 V   _8 kV_
6. 3×10^{-9} s   _3 ns_
7. 2×10^{9} Hz   _2 GHz_
8. 47×10^{-12} F   _47 pf_
9. 6.8×10^{3} Ω   _6.8 kΩ_
10. 7×10^{-3} A   _7 mA_

Practice Problems 7-B
Answers at End of Chapter

Convert the following to values with a power of 10.

1. 47 pF
2. 2 GHz
3. 3 ms
4. 8 kV
5. 4 MHz
6. 6 µs
7. 5 mH
8. 3 µV
9. 6.8 kΩ
10. 7 mA

Using the opposite procedure, a metric prefix can be converted to the corresponding power of 10, when necessary. For example,

$$9 \text{ mA} = 9 \times 10^{-3} \text{ A}$$
$$5 \text{ M}\Omega = 5 \times 10^{6} \ \Omega$$

In some cases, we may want to change from one metric prefix to another. For instance, the values in a problem may be 7 mA, 2 mA, and 800 µA. The 800 µA can be converted to 0.8 mA so that all the values are in the same units.

Table 7-2 Common Metric Prefixes Used in Electronics

Prefix	Pronunciation*	Power of 10	Example
giga	as in "gigga"	10^{9}	7 GHz = 7×10^{9} Hz
mega	as in "megga"	10^{6}	5 MΩ = 5×10^{6} Ω
kilo	as in "killoh"	10^{3}	25 kV = 25×10^{3} V
milli	as in "millie"	10^{-3}	8 mA = 8×10^{-3} A
micro	as in "mike-row"	10^{-6}	5 µH = 5×10^{-6} H
nano	as in "nannoh"	10^{-9}	3 ns = 3×10^{-9} s
pico	as in "pickoh"	10^{-12}	5 pF = 5×10^{-12} F

*Accent is always on first syllable, as in milliampere, and kilometer.

10^{-12}	10^{-9}	10^{-6}	10^{-3}	10^0	10^3	10^6	10^9	10^{12}
Pico (p)	Nano (n)	Micro (μ)	Milli (m)		Kilo (k)	Mega (M)	Giga (G)	Tera (T)

Fig. 7-1 Increasing values from left to right for the metric prefixes commonly used in electronics.

When changing between units, keep in mind the following:

$$1 \text{ M}\Omega = 1000 \text{ k}\Omega$$
$$1 \text{ k}\Omega = 0.001 \text{ M}\Omega$$

Also $1 \text{ mA} = 1000 \text{ }\mu\text{A}$
$$1 \text{ }\mu\text{A} = 0.001 \text{ mA}$$

When you go to a smaller metric unit, the coefficient should increase. Or, in converting to a larger metric unit, the coefficient should decrease.

Practice Problems 7-C
Answers at End of Chapter

1. $2000 \text{ }\mu\text{A} = $ _____ mA
2. $3 \text{ mA} = $ _____ μA
3. $0.001 \text{ mA} = $ _____ μA
4. $7 \text{ }\mu\text{A} = $ _____ mA
5. $900 \text{ k}\Omega = $ _____ MΩ
6. $0.9 \text{ M}\Omega = $ _____ kΩ
7. $5 \text{ k}\Omega = $ _____ MΩ
8. $0.002 \text{ M}\Omega = $ _____ kΩ

When a power of 10 does not correspond to the metric prefix we want to use, we can convert the power to fit the prefix, but without changing the value. For example, $70 \times 10^2 \text{ }\Omega$ can be converted to $7 \times 10^3 \text{ }\Omega$ or $7 \text{ k}\Omega$. The steps in this conversion are

$$70 \times 10^2 = 70 \times 10^2 \times \frac{10}{10} = \frac{70}{10} \times 10^3$$
$$= 7 \times 10^3 = 7 \text{ k}\Omega$$

The procedure is to multiply or divide the power to get the desired exponent, then do the opposite to the coefficient so that the value remains the same. Remember the rules for multiplying and dividing with powers of 10, as explained in Chap. 5. Note the following examples:

$$8 \times 10^4 \text{ }\Omega = 80 \times 10^3 \text{ }\Omega = 80 \text{ k}\Omega$$
$$7 \times 10^{-4} \text{ A} = 0.7 \times 10^{-3} \text{ A} = 0.7 \text{ mA}$$
$$19 \times 10^{-5} \text{ A} = 190 \times 10^{-6} \text{ A} = 190 \text{ }\mu\text{A}$$
$$22 \times 10^5 \text{ }\Omega = 2.2 \times 10^6 \text{ }\Omega = 2.2 \text{ M}\Omega$$

Practice Problems 7-D
Answers at End of Chapter

1. $8 \times 10^{-2} \text{ A} = $ _____ mA
2. $8 \times 10^{-4} \text{ A} = $ _____ mA
3. $0.6 \times 10^7 \text{ Hz} = $ _____ MHz
4. $0.6 \times 10^6 \text{ Hz} = $ _____ kHz
5. $19 \times 10^{-5} \text{ A} = $ _____ μA
6. $19 \times 10^{-7} \text{ A} = $ _____ μA
7. $3.3 \times 10^7 \text{ }\Omega = $ _____ MΩ
8. $3.3 \times 10^4 \text{ }\Omega = $ _____ kΩ

7-2 Length Units

The standard unit of length in the metric system is the meter (m), equal to 39.37 in. Originally, the meter was defined as 1×10^{-7} of the distance from the equator to the North Pole. This dimension of one ten-millionth of a quadrant of the earth's circumference was known accurately from surveying calculations about the year 1790. In 1960 the meter was redefined in terms of wavelengths of radiation of the element krypton.

Decimal multiples and fractions of the meter are listed in Table 7-3. Most common are the kilometer, which is a large value equal to 1000 m or 1×10^3 m, and the centimeter, which is only $\frac{1}{100}$ m or 1×10^{-2} m. The kilometer is pronounced "killommeter,"

Table 7-3 Metric Units of Distance

1 kilometer (km) = 10^3 meters (m)
1 hectometer (hm) = 10^2 m
1 dekameter (dam) = 10^1 m
1 centimeter (cm) = 10^{-2} m
1 millimeter (mm) = 10^{-3} m = 10^{-1} cm
1 micrometer (μm) = 10^{-6} m
1 Angstrom* (Å) = 10^{-10} m

*The Angstrom is not a metric unit.

with the accent on the second syllable. The term "micron" was formerly used in place of "micrometer," but micron is not an accepted metric term.

The most common metric–U.S. customary conversions for units of length are listed in Table 7-4. To think in metric units, a meter at 39.37 in is a little more than a yard at 36 in. The extra length is approximately 4 in, or one-third of a foot. Also, 1 meter is 3.3 ft long. A 100-m race, as an example, is 33 ft longer than the 100-yard dash.

A centimeter (cm) is a little less than one-half inch. More exactly, 1 cm = 0.394 in, which can be rounded off to 0.4 in. It takes 2.54 cm to equal 1 in. A 12-in ruler marked in centimeters has approximately 30.5 cm marks.

The kilometer (km) is used in place of the mile (mi). A distance of 1 km is equal to 0.63 mi, or almost two-thirds of a mile. The speed limit of 55 mi/h is the same as 88.55 km/h. A common metric speed limit is 90 km/h, which is equivalent to about 56.7 mi/h.

When using the conversion factors in Table 7-4, the question is when to use the form as listed and when to use its reciprocal. As given, the factor is for multiplication. Using the reciprocal is the same as dividing by the conversion factor. You can decide which to use by including all units in the calculations for a conversion. Then the units you are changing from should cancel, leaving just the unit you want. This answer shows that the correct conversion factor has been used. For instance, if you are converting from inches to centimeters, the inch units should cancel by division.

Table 7-4 Metric–U.S. Customary Conversions for Length Units

Metric Unit	English Unit	Conversion Factor
centimeter (cm)	1 in = 2.54 cm	2.54 cm/in or reciprocal*
meter (m)	1 yd = 0.9 m	0.9 m/yd or reciprocal
	1 ft = 0.3 m	0.3 m/ft or reciprocal
kilometer (km)	1 mi = 1.61 km	1.61 km/mi or reciprocal

*See text for explanation of when to use the reciprocal.

Example Convert 8 inches to centimeters.

Answer The conversion factor is 2.54 cm/in. Multiplying by 8 in, then,

$$8 \text{ in} \times \frac{2.54 \text{ cm}}{\text{in}} = 20.32 \text{ cm}$$

Notice that the inches cancel, leaving centimeters for the answer. If we used the reciprocal formula with 0.4, or 1/2.54, then

$$8 \text{ in} \times 0.4 \frac{\text{in}}{\text{cm}} = ?$$

There is no answer because none of the units can be cancelled. In the reciprocal, note that the units are inverted.

Example Convert 8 cm to inches.

Answer The conversion factor is 0.4 in/cm. Then,

$$8 \text{ cm} \times 0.4 \frac{\text{in}}{\text{cm}} = 3.2 \text{ in}$$

Practice Problems 7-E
Answers at End of Chapter

1. 12 in = ____ cm
2. 24 in = ____ cm
3. 3 ft = ____ m
4. 3 yd = ____ m
5. 3 mi = ____ km
6. 55 mi = ____ km
7. 1 m = ____ in
8. 10 km = ____ mi

7-3 Weight Units

The weight of an object is the gravitational force on its mass. One pound (lb) in the U.S. customary system of units corresponds to 2.2 kilograms (kg) in metric units. One kilogram is the weight of one thousand cubic centimeters (cm^3) of water at a specified temperature. One gram (g) is the weight of one cubic centimeter of water. In metric units, 1 kg = 1 × 10³ g. Actually, there is a platinum bar in France that represents the standard weight of one kilogram.

The most common metric units for weight, with conversion factors, are listed in Table 7-5. For a shortcut

Table 7-5 Metric–U.S. Customary Conversions for Weight Units

Metric Unit	U.S. Customary Unit	Conversion Factor
gram (g)	1 oz = 28.4 g	28.4 g/oz or reciprocal
kilogram (kg)	1 lb = 0.4545 kg	0.4545 kg/lb or reciprocal

conversion, remember that a kilogram is about two pounds. Also, a pound is about one-half kilogram. Exactly, 1 lb = 0.4545 kg.

For smaller units, there are 16 ounces (oz) in a pound. Or, 1 oz = $\frac{1}{16}$ lb. Also, a kilogram is 1000 grams. Or, 1 g = 10^{-3} kg.

The ounce is a larger unit than the gram. The exact value of 28.4 g for 1 oz in Table 7-5 is calculated by changing 1 kg to 1000 g and 1 lb to 16 oz. Starting with the fact that 1 lb is 0.4545 kg, then

$$1 \text{ oz} = 0.4545 \, \frac{\text{kg}}{\text{lb}} \times 1000 \, \frac{\text{g}}{\text{kg}} \times \frac{1}{16} \text{ lb}$$

$$= \frac{454.5}{16} \text{ g} = 28.4 \text{ g}$$

Notice that the kilograms and pounds cancel to leave just the answer of 28.4 g equal to 1 oz. Two more examples are given here now to illustrate when to multiply by the conversion factor for kilograms in Table 7-5 and when to use the reciprocal.

Example   Convert 44 lb to kilograms.

Answer   Multiply by the conversion factor in Table 7-5. Then

$$44 \text{ lb} \times 0.4545 \, \frac{\text{kg}}{\text{lb}} = 20 \text{ kg (approx)}$$

Example   Convert 20 kg to pounds.

Answer   Use the reciprocal, with 2.2 for 1/0.4545 and lb/kg instead of kg/lb. Then

$$20 \text{ kg} \times 2.2 \, \frac{\text{lb}}{\text{kg}} = 44 \text{ lb}$$

Both of these answers check out with the fact that 1 kg = 2.2 lb. Note that the reciprocal is still used as a multiplying factor in the conversion. Multiplying by a reciprocal $1/N$ is the same as dividing by N.

Practice Problems 7-F
Answers at End of Chapter

1. 4.4 lb = ____ kg
2. 10 lb = ____ kg
3. 44 lb = ____ kg
4. 57 lb = ____ kg
5. 8 oz = ____ g
6. 16 oz = ____ g
7. 19.7 g = ____ oz
8. 9.92 kg = ____ lb

7-4 Volume Units

Volume is a space measured in three dimensions. Consider the cube in Fig. 7-2, where each side has a length of 10 cm. The product of two sides corresponds to an area of 10 cm × 10 cm = 100 cm². Give the area a height of 10 cm and the resulting volume is 100 cm² × 10 cm = 1000 cm³. Note that the centimeter units of length are multiplied along with the numerical values. This cube of 1000 cm³ contains 1000 smaller cubes, each with a volume of 1 cm³. In the metric system, 1000 cm³ is the standard volume of one liter (L). The volume of 1 cm³ is 1×10^{-3} L or 1 milliliter (mL).

The two main conversions between metric and U.S. customary units of volume are listed in Table 7-6. For a shortcut conversion, remember that one liter is about

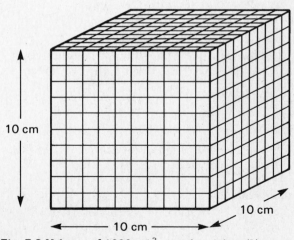

Fig. 7-2 Volume of 1000 cm³, equal to 1 liter (L).

Table 7-6 Metric–U.S. Customary Unit Conversions for Volume

Metric Unit	U.S. Customary Unit	Conversion Factor
liter (L) = 1000 cm^3	1 quart (qt) = 0.95 L	0.95 L/qt or reciprocal
cubic centimeter (cm^3)	1 in^3 = 16.4 cm^3	16.4 cm^3/in^3 or reciprocal

the same as one quart (qt). Specifically, the quart is slightly less, with 1 qt being equal to 0.95 L. A gallon, equal to 4 qt, is about the same as 4 L.

The conversion between cubic centimeters and cubic inches in Table 7-6 is derived as follows:

$$1 \text{ in} = 2.54 \text{ cm}$$
$$1 \text{ in}^3 = (2.54 \text{ cm})^3 = 16.39 \text{ cm}^3$$
$$= 16.4 \text{ cm}^3$$

In U.S. customary units, we also have a pint, equal to one-half quart. The pint, then, is also about one-half liter.

In addition, there are smaller volume units: 16 fluid ounces to a pint, or 32 fluid ounces to a quart. A tablespoon of liquid corresponds to the volume of one-half a fluid ounce. Two tablespoons equal one fluid ounce, which equals the volume of 30 cm^3 or 30 mL.

This variety of different volume units in the U.S. customary system seems especially complicated compared with the liters and millimeters in the metric system.

Practice Problems 7-G
Answers at End of Chapter

Give exact answer.

1. 4 qt = _____ L
2. 10 qt = _____ L
3. 500 mL = _____ L
4. 17 cm^3 = _____ mL
5. 10 in^3 = _____ cm^3
6. 10 in^3 = _____ mL
7. 3 mL = _____ cm^3
8. 5 L = _____ qt

7-5 Metric Units of Power, Work, and Energy

It should be noted that work and energy have the same units. Power, though, is the time rate of doing work, or power is work divided by time. All these units are derived from the units of force. The force units are based on Newton's law of acceleration (a) acting on a mass (M); the formula is

$$F = Ma$$

A force tends to produce motion. The greater the force, the faster the motion is for the same mass. Acceleration is a time rate of change in the velocity of the motion.

Velocity and speed have the same units, but velocity also has direction and is called a *vector* quantity. Velocity is a time rate of change in distance. As an example, the velocity can be 3 cm/s. If this velocity is increased to 4 cm/s, the change in velocity is the acceleration. This value of acceleration is 1 cm/s/s or 1 cm/s^2. Read this as one centimeter per second per second.

For a mass M of 1 g and acceleration a of 1 cm/s^2, the force F is a *dyne* (dyn). A force of a dyne on a 1-g mass will make it move with an acceleration of 1 cm/s^2. The dyne is the basic unit of force in the centimeter-gram-second (cgs) system of metric units.

For a mass M of 1 kg and acceleration a of 1 m/s^2, the force F is 1 *newton* (N). This unit is named after Isaac Newton (1642–1727), the famous scientist who discovered the laws of motion. A force of 1 N on a 1-kg mass will make it move with an acceleration of 1 m/s^2. The newton is the basic unit of force in the meter-kilogram-second (mks) system of units.

The mks units are larger than cgs units, making them a practical system. The Système International (SI) units, standardized for use throughout the world, are essentially the same as mks units. They are used in science, engineering, and industry for the units in mechanics, electricity, electronics, light, and heat.

A comparison between the cgs units and the larger SI units is given in Table 7-7. The main SI units to remember are the newton (N) for force, the newton-meter or joule (J) for work or energy, and the watt (W) for power, equal to 1 J/s.

In the U.S. customary system of foot-pound-second (fps) units, common units for power are foot-pounds

Table 7-7 Units of Force, Power, Work, and Energy

Type	CGS Unit	SI Unit	Conversion Factor
Force	dyne (dyn)	1 newton (N) = 10^5 dyn 1 N = 1 kg·m/s²	10^5 dyn/N or reciprocal
Work or energy	dyn·cm = erg	1 joule (J) = 1 N·m = 10^7 erg	10^7 erg/J or reciprocal
Power	erg/s	1 watt (W) = 1 J/s = 10^7 erg/s	10^7 erg/s/W or reciprocal

per second (ft·lb/s) and horsepower (hp). The unit of 1 hp is equal to 550 ft·lb. For conversion to metric units, 1 hp = 746 W.

Practice Problems 7-H
Answers at End of Chapter

1. 5 kg·m/s² = _____ N
2. 5 N·m = _____ J
3. 5 J/s = _____ W
4. 1100 ft·lb = _____ hp
5. 2 hp = _____ W
6. 14 J/s = _____ W
7. 53 W·s = _____ J
8. 37 J/2 s = _____ W

7-6 Temperature and Heat Units

The temperature units marked on a thermometer indicate the relative amount of heat energy in an object. Different temperature scales can be used, depending on the reference for the heat comparison. However, the actual amount of heat energy is measured in calories or joules.

Metric units for temperature are given using the Celsius scale, formerly known as the centigrade scale, indicated as °C. On this scale, 0° is the freezing point of water and 100° is the boiling point.

In U.S. customary units, the Fahrenheit scale, invented by G. D. Fahrenheit, is still used for weather observations and general purposes. On this scale, the freezing point of water is taken as 32° and the boiling point as 212°. To convert from one scale to the other, remember that the Fahrenheit scale has 180 divisions between freezing and boiling and the Celsius scale has 100 divisions between freezing and boiling. This comparison is the reason why the conversion formulas use the ratio $^{100}/_{180}$, or $^5/_9$, and its reciprocal value of $^9/_5$.

For °C,

$$T_C = \frac{5}{9}(T_F - 32°) \qquad (7\text{-}1)$$

For °F,

$$T_F = \left(\frac{9}{5}T_C\right) + 32° \qquad (7\text{-}2)$$

To check Formula (7-2), we can convert 212°F to the Celsius scale:

$$T_C = \frac{5}{9}(212° - 32°)$$

$$= \frac{5}{9}(180) = \frac{900}{9}$$

$$T_C = 100°$$

The boiling point of water is 212°F or 100°C.

For the opposite conversion, change 0°C to the Fahrenheit scale. Then

$$T_F = \left(\frac{9}{5} \times 0\right) + 32°$$

$$= 0 + 32°$$

$$T_F = 32°$$

The freezing point of water is 0°C or 32°F.

When making these temperature conversions, remember that T_C must be a smaller number than the equivalent T_F. Therefore, the multiplying factor is $^5/_9$. Also, the 32° is subtracted. The 32° must be subtracted before you multiply by $^5/_9$.

For the opposite conversion, T_F must be a larger number than the equivalent T_C. Then the multiplying factor is $^9/_5$. Also, the 32° is added. However, the T_C is multiplied by $^9/_5$ before you add 32°.

In the metric system, our body temperature is exactly 37°C. Average room temperature is 20°C. A cold day (temperatures below freezing) would have a negative Celsius temperature.

A third temperature scale is the absolute or kelvin scale. On this scale, the zero point is absolute zero, −273°C. At 0 K, any material loses all its heat energy. Note that the symbol is just K, without the degree mark. This scale, devised by Lord Kelvin, is used for the SI unit of temperature.

To convert from degrees Celsius to Kelvins, just add 273°. As examples, 0°C is 273 K and 20°C = 293 K. For the opposite case, subtract 273 to change from Kelvins to degrees Celsius. For instance, 0 K equals −273°C.

Practice Problems 7-I
Answers at End of Chapter

1. 70°F = ___ °C
2. 90°F = ___ °C
3. 10°F = ___ °C
4. 25°C = ___ °F
5. 0°C = ___ K
6. 100°C = ___ K
7. 20°C = ___ °F
8. 37°C = ___ °F

The metric unit for heat energy is the calorie (cal). One calorie is the heat energy needed to raise the temperature of one gram of water by one degree Celsius. The mks unit is the kilocalorie, equal to 1×10^3 cal. In the SI system, though, the standard unit of heat energy is the joule unit of work or energy, where

$$1 \text{ cal} = 4.19 \text{ J}$$

or

$$1 \text{ J} = 0.24 \text{ cal}$$

In the U.S. customary foot-pound-second system, the unit of heat energy is the British thermal unit (Btu), where

$$1 \text{ Btu} = 252 \text{ cal}$$

or

$$1 \text{ Btu} = 1056 \text{ J}$$

All these values are approximate, as 1 cal is exactly 4.184 J.

Practice Problems 7-J
Answers at End of Chapter

1. 100 cal = ___ J
2. 419 J = ___ cal
3. 10 Btu = ___ cal
4. 2520 cal = ___ Btu
5. 756 cal = ___ J
6. 2 Btu = ___ cal
7. 504 cal = ___ J
8. 3137 cal = ___ J

Review Problems
Answers to Odd-Numbered Problems at Back of Book

These problems summarize operations with metric units and prefixes.

1. 0.0024 A = ___ mA
2. 2.4 mA = ___ μA
3. 29,000 V = ___ kV
4. 800 kHz = ___ MHz
5. 0.000 022 μF = ___ pF
6. 78×10^{-4} A = ___ mA
7. 36 in = ___ cm
8. 50 mi = ___ km
9. 10 lb = ___ kg
10. 10 oz = ___ g
11. 10 qt = ___ L
12. 10 gal = ___ L
13. 20 in³ = ___ cm³
14. 5 N·m = ___ J
15. 44 J/s = ___ W
16. 17 W·s = ___ J
17. 80°F = ___ °C
18. 26.7°C = ___ K
19. 100 cal = ___ J
20. 5 Btu = ___ J

Answers to Practice Problems

7-A
1. 9 mA
2. 5 MΩ
3. 6 μs
4. 4 MHz
5. 8 kV
6. 3 ns
7. 2 GHz
8. 47 pF
9. 6.8 kΩ
10. 7 mA

7-B
1. 47×10^{-12} F
2. 2×10^9 Hz
3. 3×10^{-3} s
4. 8×10^3 V
5. 4×10^6 Hz
6. 6×10^{-6} s
7. 5×10^{-3} H
8. 3×10^{-6} V
9. 6.8×10^3 Ω
10. 7×10^{-3} A

7-C
1. 2 mA
2. 3000 μA
3. 1 μA
4. 0.007 mA
5. 0.9 MΩ
6. 900 kΩ
7. 0.005 MΩ
8. 2 kΩ

7-D
1. 80 mA

2. 0.8 mA
3. 6 MHz
4. 600 kHz
5. 190 μA
6. 1.9 μA
7. 33 MΩ
8. 33 kΩ

7-E
1. 30.48 cm
2. 60.96 cm
3. 0.9 m
4. 2.7 m
5. 4.83 km
6. 88.55 km
7. 39.37 in
8. 6.2 mi

7-F
1. 2 kg
2. 4.545 kg
3. 20 kg
4. 25.9 kg
5. 227.2 g
6. 454.4 g
7. 0.7 oz
8. 21.8 lb

7-G
1. 3.8 L
2. 9.5 L
3. 0.5 L
4. 17 mL
5. 164 cm³
6. 164 mL

7. 3 cm³
8. 5.26 qt

7-H
1. 5 N
2. 5 J
3. 5 W
4. 2 hp
5. 1492 W
6. 14 W
7. 53 J
8. 18.5 W

7-I
1. 21.1°C
2. 32.2°C
3. −12.2°C
4. 77°F
5. 273 K
6. 373 K
7. 68°F
8. 98°F

7-J
1. 419 J
2. 100 cal
3. 2520 cal
4. 10 Btu
5. 3168 J
6. 504 cal
7. 2112 J
8. 13,144 J

8 ALGEBRA

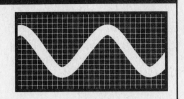

Algebra is a branch of mathematics that deals with the ordinary arithmetic calculations covered in Chap. 1 but uses letter symbols in addition to numbers. The purpose is to make general statements about values in the form of an equation. For instance, by means of the formula $V = IR$, we can find the voltage V for any values of I and R.

Remember that the letters represent not decimal places but entire quantities. The symbol V can have the value of 7, 70, or 700, for example. Also, letters written together indicate not decimal places but multiplication. IR means I times R. It is the same as $I \times R$ or $I \cdot R$ or $(I)(R)$.

More details are explained in the following sections:

8-1   Literal Numbers
8-2   Adding or Subtracting Literal Numbers
8-3   Powers and Roots of Literal Numbers
8-4   Multiplying or Dividing Literal Numbers with Exponents
8-5   Fractions with Literal Numbers
8-6   Terms and Factors
8-7   Polynomials
8-8   Cancelling Literal Numbers in a Fraction
8-9   Binomials
8-10   Factoring

8-1 Literal Numbers

Letters such as a, b, c, x, y, and z are used as literal numbers in algebra to represent the decimal values of numerical counting. For instance, a can represent 3, and $2a$ is then 6. The purpose of literal numbers is to state the relationship between quantities in a general form. Letters at the beginning of the alphabet, such as a, b, and c, usually represent constant or known values, while x, y, and z are used for unknown values. Either lowercase letters or capitals can be used.

Remember that for a literal number such as $3a$, the letter represents a complete numerical value, not just the count in a decimal place. The proof is that ab is the same as ba. The order of letters does not affect the value. Some examples here illustrate how literal numbers are used.

ab is $a \times b$ $a \times a$ is a^2
$2a$ is $2 \times a$ $\sqrt{a^2}$ is a
$a + a$ is $2a$ $1/a$ is $1 \div a$

Note that when a letter has a numerical coefficient, as in $2a$, the number is written first.

Practice Problems 8-A
Answers at End of Chapter

Answer true or false for the following.

1. $6ab = 6 \times a \times b$ *thu*   6. $xy = (x)(y)$
2. $c^3 = c \times c \times c$ 7. $cb = c + d$
3. $y^2 = y \times y$ 8. $a^0 = 1$
4. $1/y = 1 \div y$ 9. $a/b = a \div b$
5. $-2a = 2 - a$ 10. $ab = a \times b$

8-2 Adding or Subtracting Literal Numbers

Only the same letters can be added or subtracted.

> **Examples** $5a + 2a = 7a$
> $5a - 2a = 3a$
> $5a + 2a + 2b = 7a + 2b$

In the last example, note that the unlike terms $7a$ and $2b$ cannot be combined. Identical letters with different subscripts indicate different literal numbers. To give an example, V_1 and V_2 represent two different quantities.

Identical letters with different exponents also indicate different literal numbers. As an example, a and a^2 are two different literal numbers, which therefore cannot be combined by addition or subtraction.

Practice Problems 8-B
Answers at End of Chapter

Add or subtract the following literal numbers.

1. $x + x =$ 3. $5a - 3a =$
2. $5a + 3a =$ 4. $6y + y =$

5. $5y + 4y =$
6. $2a + 3c + 3a =$
7. $6y - y =$
8. $V_1 + 5V_1 =$
9. $R_1 + 4R_3 + 2R_1 =$
10. $I_1 + 6I_1 =$

8-3 Powers and Roots of Literal Numbers

An exponent is used in the same way with a literal number as with a decimal number. For instance, a^2 is equal to $a \times a$. Also, a^3 is $a \times a \times a$. Here the literal number a is the base raised to a higher power.

When the letter has a numerical coefficient included with the literal base, both are raised to the higher power.

Example Square $3a$.

Answer This is written as $(3a)^2$. Each factor is squared, as

$$3^2 \times a^2 = 9a^2$$

Or the problem can be stated as

$$3a \times 3a = 9a^2$$

If the number were written as $3a^2$, then it would be understood that only the a was squared. Suppose $a = 4$. Then

$$3a^2 = 3 \times 4^2 = 3 \times 16 = 48$$

For $(3a)^2$, however, the value is $(12)^2$, which equals 144.

When an exponent of a literal number is raised to any power, the two exponents are multiplied.

Example Cube the value of (a^2).

Answer This is written $(a^2)^3$.
Since the exponents are multiplied, $2 \times 3 = 6$. Therefore,

$$(a^2)^3 = a^6$$

As with decimal numbers, any literal number to the zero power is 1. For example, $x^0 = 1$.

Any literal number to the 1 power is the literal number itself. For example, $a^1 = a$ and $(xy)^1 = xy$.

Practice Problems 8-C
Answers at End of Chapter

Raise to the indicated power.

1. $(2b)^3 =$
2. $(4y)^2 =$
3. $(5x)^3 =$
4. $(a^5)^3 =$
5. $(3c)^2 =$
6. $(2z)^4 =$
7. $(-3a)^2 =$
8. $(a^0)^3 =$
9. $(5a^2)^2 =$
10. $(2x)^3 =$

A root can also be found for literal numbers. The procedure is the same as with decimal numbers.

Example Find the fourth root of a^8.

Answer This is written as $\sqrt[4]{a^8}$. Divide the exponent by the root: $8/4 = 2$. Therefore,

$$\sqrt[4]{a^8} = a^2$$

It should be noted that $\sqrt{a^2}$ is really $\pm a$, as either a or $-a$ squared results in a^2. The problem usually will provide a clue as to whether the answer is $+$ or $-$ or both.

For reciprocals, the rules for literal numbers are the same as those for powers of 10. Change the sign of the exponent when moving the number between denominator and numerator. As examples,

$$\frac{1}{a^2} = a^{-2} \qquad \text{or} \qquad \frac{1}{a^{-2}} = a^2$$

Here the base number is a and its exponent is either 2 or -2.

Practice Problems 8-D
Answers at End of Chapter

Find the indicated root.

1. $\sqrt[3]{8b^3} =$
2. $\sqrt{16y^2} =$
3. $\sqrt{a^6} =$
4. $(a^6)^{1/3} =$
5. $\sqrt{25y^4} =$
6. $\sqrt{a^{14}} =$
7. $\sqrt{10a^2} =$
8. $\sqrt[3]{-8a^3} =$
9. $\sqrt[3]{8a^6} =$
10. $(y^8)^{1/2} =$

Practice Problems 8-E
Answers at End of Chapter

Assuming a is 2 and b is 3, evaluate.

1. $(2b)^2 =$
2. $2b^2 =$
3. $4a^2 + (2a)^2 =$
4. $\sqrt[3]{8b^3} =$
5. $(8b^3)^{1/3} =$
6. $\sqrt{16a^6} =$
7. $(ab)^2 =$
8. $ab^2 =$
9. $a^2b =$
10. $ab =$

8-4 Multiplying or Dividing Literal Numbers with Exponents

Multiply the numerical coefficients of the literal base to obtain the new numerical coefficient, but add the exponents to obtain the new exponent of the base. However, the exponents can be combined this way only if the base is the same for both numbers.

Example Multiply $6a^2$ by $3a^3$.

Answer The problem is written $6a^2 \times 3a^3$. Multiply the coefficients:

$$6 \times 3 = 18$$

Since the base of these exponents is the same for both terms, the exponents are added:

$$2 + 3 = 5$$

The final answer is, then,

$$6a^2 \times 3a^3 = 18a^5$$

When no exponent is shown for the base, it actually is 1. For instance, a is the same as a^1. Also, the numerical coefficient is 1 if none is shown; a^3 is the same as $1a^3$.

Practice Problems 8-F
Answers at End of Chapter

Multiply the following.

1. $y \times y =$
2. $2y \times 3y =$
3. $4a \times 2a^2 =$
4. $4a^2 \times 2a =$

5. $2a^5 \times 3a^4 =$
6. $3x^2 \times 2x^5 =$
7. $3b \times 5b^3 =$
8. $2I \times 4I =$
9. $3a \times 2b =$
10. $3x^a \times 2x^b =$

To divide literal numbers, divide the numerical coefficients of the literal base to find the new coefficient, and subtract the exponent of the divisor from the exponent of the dividend to obtain the new exponent. Again, the exponents can be combined only for the same base.

Example Divide $8a^3$ by $2a$.

Answer The problem is written as

$$8a^3 \div 2a = \frac{8a^3}{2a}$$

The divisor is $2a$. Divide the coefficients: $8 \div 2 = 4$. Then subtract the exponent of the divisor (1 in this case) from the exponent of the dividend.

$$3 - 1 = 2$$

Therefore,

$$\frac{8a^3}{2a} = 4a^2$$

Practice Problems 8-G
Answers at End of Chapter

Divide the following.

1. $8a^6 \div 4a^4 =$
2. $12a^3 \div 4a =$
3. $7x^7 \div 2x^2 =$
4. $12b^6 \div 6b^2 =$
5. $8a \div 2b =$
6. $8a^4 \div 4a^7 =$
7. $3y \div 3y =$
8. $3x^a \div 2x^b =$

8-5 Fractions with Literal Numbers

Literal numbers represent entire quantities that could be part of fractions. Thus, a literal number can be in the numerator or the denominator, for example, $a/2$, $2x/3$, x/y, $(a + b)/c$. Remember also that the *value* of the literal number may in fact be a fraction itself.

Fractions containing literal numbers can be added, subtracted, multiplied, and divided just as decimal numbers can. They can also be raised to a power and have roots extracted.

To add or subtract fractions, the denominators must be the same, whether literal numbers or regular numbers. Once the denominators are the same, the numerators are combined as in ordinary addition and subtraction.

Example   Add $2/a$ and $3/a$.

Answer $\dfrac{2}{a} + \dfrac{3}{a} = \dfrac{2+3}{a} = \dfrac{5}{a}$

To multiply fractions containing literal numbers, multiply the numerators together to obtain the new numerator and multiply the denominators together to obtain the new denominator.

Example   Multiply $2/a \times 3/a$.

Answer $\dfrac{2}{a} \times \dfrac{3}{a} = \dfrac{6}{a^2}$

To divide fractions, invert the divisor and multiply.

Example   Divide $2/a$ by $3/b$.

Answer $\dfrac{2}{a} \div \dfrac{3}{b} = \dfrac{2}{a} \times \dfrac{b}{3} = \dfrac{2b}{3a}$

To raise a fraction to a power, both the numerator and denominator are raised to that power.

Example   Find the cube of $a/2$.

Answer $\left(\dfrac{a}{2}\right)^3 = \dfrac{a^3}{2^3} = \dfrac{a^3}{8}$

To find the root of a fraction, find the root of both numerator and denominator.

Example   Find the cube root of $a^3/8$.

Answer $\sqrt[3]{\dfrac{a^3}{8}} = \dfrac{\sqrt[3]{a^3}}{\sqrt[3]{8}} = \dfrac{a}{2}$

Practice Problems 8-H
Answers at End of Chapter

Do the following problems with combined operations.

1. $\dfrac{x}{5} + \dfrac{x}{5} =$

2. $\dfrac{2a}{3} - \dfrac{a}{3} =$

3. $\dfrac{2a+1}{2} + \dfrac{2a-1}{4} =$

4. $\left(\dfrac{x}{3}\right)\left(\dfrac{y}{2}\right) =$

5. $\left(\dfrac{2}{3}\right)\left(\dfrac{a}{b}\right) =$

6. $\dfrac{a}{b} \div \dfrac{a}{b} =$

7. $\dfrac{a}{2} \times \dfrac{ab}{4} =$

8. $\dfrac{xy}{a} \div \dfrac{a}{xy} =$

9. $\dfrac{a^2b}{a} \times \dfrac{cd}{ab} =$

10. $\left(\dfrac{2}{b}\right)^2 =$

11. $\left(\dfrac{a}{b}\right)^3 =$

12. $\sqrt{\dfrac{a^2}{b^2}} =$

13. $\sqrt{4a^2} \times 2a =$

14. $\sqrt{\dfrac{4}{b^2}} =$

8-6 Terms and Factors

Factors are numbers to be multiplied or divided. Terms are numbers that are added or subtracted. For example, in the expression 3 + 2, the 3 and 2 are terms. In the expression $3a + 2a$, the $3a$ and $2a$ are terms, but 3 and a are factors in the term $3a$. The number 3 is further described as the coefficient of the term.

If we consider the expression (3)(2), the 3 and 2 are *factors* of the product 6. This is not the same as the decimal count of 32.

It is important to realize the fundamental difference between factors and terms, because they have different rules of operation in algebraic equations. For instance, in $x = 2 + 3$, all the numbers are terms without any factors. However, in $V = IR$, the I and R are factors of V. The details of working with such equations are explained in Chap. 9.

Practice Problems 8-I
Answers at End of Chapter

Pick out the terms in the following expressions.

1.	$2x + 3$	**4.**	$xy^2 + 8$	**7.**	$ab + xy$
2.	$5y - 4$	**5.**	$R_1 + R_2$	**8.**	$2 + ab$
3.	$abc + a$	**6.**	$C_1 + C_2$		

Pratice Problems 8-J
Answers at End of Chapter

Pick out the factors in the following expressions.

1.	$2x$	**4.**	xy^2	**7.**	IR
2.	$5y$	**5.**	R_1R_2	**8.**	VI
3.	abc	**6.**	$2\pi fL$		

Practice Problems 8-K
Answers at End of Chapter

Evaluate the following when $x = 4$ and $y = 5$.

1.	$xy =$	**4.**	$x^2 + y =$	**7.**	$2x + y =$
2.	$x + y =$	**5.**	$y^2x =$	**8.**	$2y + x =$
3.	$x^2y =$	**6.**	$y^2 + x =$		

8-7 Polynomials

A *polynomial* is an algebraic expression using literal numbers with more than one term. When only one term is used, it is a *monomial*. Examples of a monomial are a, $2a$, $3x$, $5y^2$, and $4ab$.

A *binomial* has two terms. Examples are $(a + b)$ and $(3x + 5y^2)$. Binomials with the same terms but opposite signs are called *conjugates*. For instance, $(a + b)$ and $(a - b)$ are conjugate binomials.

A *trinomial* has three terms. An example is $a^2 + 2ab + b^2$. Trinomials and binomials are also polynomials. A polynomial can also have more than three terms.

Example Identify each of the following as a monomial, binomial, or trinomial.

$$a$$
$$a + b$$
$$a^2 + b^2$$
$$ab$$
$$a^3 + bc$$
$$a + b + c$$
$$a^2 + b^2 + c^2$$
$$2a^2 + 3b^2 + 4ab$$
$$3a^2b^2c^3$$

Answer a is a monomial (one term: a).
$a + b$ is a binomial (two terms: a and b).
$a^2 + b^2$ is a binomial (two terms: a^2 and b^2).
ab is a monomial (one term: ab).
$a^3 + bc$ is a binomial (two terms: a^3 and bc).
$a + b + c$ is a trinomial (three terms: a, b, and c).
$a^2 + b^2 + c^2$ is a trinomial (three terms: a^2, b^2, and c^2).
$2a^2 + 3b^2 + 4ab$ is a trinomial (three terms: $2a^2$, $3b^2$, and $4ab$).
$3a^2b^2c^3$ is a monomial (one term: $3a^2b^2c^3$).

With polynomials, it is general practice to write the terms in descending order of exponents for one of the common letters. Some examples are

$$x^3 + 5x^2 + x$$
$$2a^3 + 3a^2 + 6$$
$$x^3y + x^2y^2 + xy$$

To multiply a polynomial by a monomial, multiply each term in the polynomial by the monomial.

Example Multiply $x + y$ by $2xy$.

Answer The problem can be written as

$$2xy(x + y)$$

Both terms in the binomial $(x + y)$ are multiplied by $2xy$ to produce a new binomial:

$$(2xy)(x) + (2xy)(y)$$

or

$$2x^2y + 2xy^2$$

Therefore,

$$2xy(x + y) = 2x^2y + 2xy^2$$

To divide a polynomial by a monomial, divide each term in the polynomial by the monomial.

Example Divide $6x^2 + 12x$ by $3x$

Answer The problem can be written as

$$\frac{6x^2 + 12x}{3x} = \frac{6x^2}{3x} + \frac{12x}{3x}$$

$$= 2x + 4$$

To multiply two polynomials, multiply each term in one polynomial by each term in the other polynomial and add the partial products.

Example Multiply $x^2 + x + 3$ by $x - 1$.

Answer The problem can be written similar to the way a numerical multiplication problem is done:

$$\begin{array}{r} x^2 + x + 3 \\ x - 1 \\ \hline \end{array}$$

Now proceed in a fashion similar to multiplication in arithmetic, but start with the left-hand term. The following multiplications are done and the partial products are recorded as shown:

$$x \times x^2 = x^3 \qquad -1 \times x^2 = -x^2$$
$$x \times x = x^2 \qquad -1 \times x = -x$$
$$x \times 3 = 3x \qquad -1 \times 3 = -3$$

Multiplying the two polynomials now,

$$\begin{array}{r} x^2 + x + 3 \\ x - 1 \\ \hline x^3 + x^2 + 3x \\ -x^2 - x - 3 \\ \hline x^3 + 0 + 2x - 3 \end{array}$$

The answer, then, is $x^3 + 2x - 3$

Note that the partial products are added with like terms in line.

The methods of dividing polynomials are similar to those performed in arithmetic, but they are not shown here because they seldom apply to electronics problems. More details can be found in algebra textbooks.

Practice Problems 8-L
Answers at End of Chapter

Do the operations indicated.

1. $a^2 \times 2ab =$
2. $a^2 \times (2ab + 4) =$
3. $a^2 \times (2ab - 4) =$
4. $(2a^3b + 4a^2) \div a^2 =$
5. $4abc \div ab =$
6. $(2y + 1) \times y =$
7. $(x + y)^2 =$
8. $(x + y)(x - y) =$
9. $(x - y)^2 =$
10. $8xyz \div 4yz =$

8-8 Cancelling Literal Numbers in a Fraction

Whether literal numbers can be cancelled depends on whether we are considering terms or factors. An example with factors is

$$\frac{abc}{a} = bc$$

The same factor of a in numerator and denominator can be cancelled here. This cancellation corresponds to the numerical example

$$\frac{2 \times 3 \times 4}{2} = 3 \times 4 = 12$$

However, consider the following example with terms instead of factors:

$$\frac{a + b + c}{a} = \;?$$

The a in the numerator here cannot be cancelled because it is a term, not a factor.

The following expression also has terms, but each term has factors:

$$\frac{a + ab + ac}{a} = 1 + b + c$$

Here the a can be cancelled, but only because it is in *all the terms*. Note that the 1 in the answer comes from a/a, as a is actually $1a$.

The idea here is that the fraction bar is a sign of grouping. All the terms in the numerator are divided by a in the denominator. You cannot cancel a factor in any term unless it is cancelled in all the terms.

It should be noted that a factor can include a group of terms. Note the following examples:

$$\frac{(x + y)(a + b)}{c(a + b)} = \frac{x + y}{c}$$

The $(a + b)$ can be cancelled because it is a factor of $(x + y)$ in the numerator and c in the denominator.

Practice Problems 8-M
Answers at End of Chapter

Simplify these fractions by cancellation, if possible.

1. $\dfrac{8b}{2} =$ 5. $\dfrac{xyz}{z} =$ 9. $\dfrac{4ab}{2a} =$

2. $\dfrac{ab}{a} =$ 6. $\dfrac{ab + a}{a} =$ 10. $\dfrac{4ab}{2b} =$

3. $\dfrac{a + b}{a} =$ 7. $\dfrac{a^2b}{a} =$

4. $\dfrac{abcd}{ac} =$ 8. $\dfrac{3b^2c}{bc} =$

8-9 Binomials

A binomial as defined in Sec. 8-7, has two terms. Each term can be a literal or decimal number or a combination of both. An example of a binomial we can use here is $(x + 3)$. There are two special cases of such binomials that can be useful in simplifying algebraic expressions. One example is the product of conjugate binomials, such as $(x + 3)(x - 3)$. The other example is the square of any binomial, such as $(x + 3)^2$ or $(x - 3)^2$.

Conjugate binomials have the same terms with opposite signs. The useful feature is that their product is always equal to another binomial with the difference of the squares of the two terms. As a formula,

$$(x + a)(x - a) = x^2 - a^2$$

For our example of $(x + 3)(x - 3)$, the product is $x^2 - 9$. To prove it, we can do this problem by longhand multiplication. Then

$$
\begin{array}{r}
x + 3 \\
x - 3 \\
\hline
x^2 + 3x \\
- 3x - 9 \\
\hline
x^2 \pm 0 - 9 \\
\end{array}
$$
partial products
$= x^2 - 9$ (answer)

Note that the middle terms with x in the partial products cancel because they are equal and opposite. The x^2 term is positive because it is the product of two positive values. The -9 for the last term is negative because it is the product of two terms with opposite signs.

The simplification here is to take a binomial with squares in the terms and divide it into simpler binomials without squares. For this example, $x^2 - 9$ is $(x + 3)(x - 3)$.

Another special case is the square of a binomial, as in $(x + 3)^2$. The longhand multiplication is

$$
\begin{array}{r}
x + 3 \\
x + 3 \\
\hline
x^2 + 3x \\
+ 3x + 9 \\
\hline
x^2 + 6x + 9 \quad \text{(answer)}
\end{array}
$$
} partial products

The answer is a trinomial. This type is considered a perfect trinomial square because it is exactly equal to the square of a binomial.

A shorthand method of squaring the binomial is as follows:

1. Square the first term in the binomial for the first term in the trinomial.
2. Square the other term in the binomial for the last term in the trinomial.
3. Multiply the first term by the second term of the binomial and double the result. This is the middle term in the trinomial.

We can apply these rules to the example of $(x - 3)^2$. Then

1. x^2 is the first term in the trinomial.
2. 9, equal to $(-3)^2$, is the last term in the trinomial.
3. $(x) \times (-3)$ doubled is $-6x$ for the middle term in the trinomial.

The final result for the square of the binomial is

$$(x - 3)^2 = x^2 - 6x + 9$$

The simplification in this example involves taking a perfect trinomial square and dividing it into two simpler binomials. For this example, $x^2 - 6x + 9$ is $(x - 3)(x - 3)$.

Practice Problems 8-N
Answers at End of Chapter

1. $(x + 3)(x - 3) =$
2. $(x + 3)^2 =$
3. $(x - 3)^2 =$
4. $(3a + 2b)(3a - 2b) =$

5. $x^2 - 4 = (\quad)(\quad)$
6. $x^2 + 6x + 9 = (\quad)^2$
7. $x^2 - 6x + 9 = (\quad)^2$
8. $9a^2 - 4b^2 = (\quad)(\quad)$

8-10 Factoring

Factors of a number are the parts that can be multiplied to produce the original number. For example, 2 and 4 are factors of 8, as $2 \times 4 = 8$, and a and b are factors of ab, since $a \times b = ab$. Factoring is often helpful with literal numbers in algebra to find parts that can be cancelled to simplify the expression.

One method is to factor out one or more letters in a group of terms. For instance,

$$ayx^2 + ayx = ay(x^2 + x)$$

Note that the common factor ay must be exactly the same in all the terms. Otherwise it cannot be factored out.

A useful method is to convert the difference between two squares into conjugate binomials when they can be used for cancellation. As an example,

$$\frac{a^2 - b^2}{a + b} = \frac{(a - b)(a + b)}{(a + b)} = a - b$$

Another technique is to convert a perfect trinomial square into the square of a binomial, where the binomial can be used for cancellation. As an example,

$$\frac{x^2 + 6x + 9}{x + 3} = \frac{(x + 3)(x + 3)}{x + 3} = x + 3$$

The trinomial here is a perfect square, as $(x + 3)^2 = x^2 + 6x + 9$.

Practice Problems 8-O
Answers at End of Chapter

Simplify the following expressions by factoring and cancellation, if possible.

1. $\dfrac{8x^2 + 8x}{8}$

2. $\dfrac{3b^2x^2 + 2b^2x}{b^2}$

3. $\dfrac{2axy + 2bxy}{2xy}$

4. $\dfrac{6x^3 + 18x}{2x}$

5. $\dfrac{(x + y)(x - y)}{x - y}$ **7.** $\dfrac{(4a + 3)^2}{4a + 3}$

6. $\dfrac{x^2 - y^2}{x - y}$ **8.** $\dfrac{16a^2 + 24a + 9}{4a + 3}$

Review Problems
Answers to Odd-Numbered Problems at Back of Book

The following problems summarize the algebraic methods explained in this chapter.

1. $4a^2 + 2a^2 =$ **3.** $(3xy)^2 =$

2. $6ab - ab =$ **4.** $\sqrt{9a^2b^2} =$

5. $3a^2 \times 2a =$ **12.** $\dfrac{6a^2 + 3a^2}{3a} =$

6. $3a^2 \div 2a =$ **13.** $2y(y + 3) =$

7. $\dfrac{a}{5} + \dfrac{3a}{5} =$ **14.** $\dfrac{4y^3 + 2y^2}{2y} =$

8. $\dfrac{2a}{3} \times \dfrac{7a^2}{5} =$ **15.** $(3a + 2b)^2 =$

9. $\dfrac{5a^2}{3} \div \dfrac{5}{3a} =$ **16.** $(3a - 2b)^2 =$

10. $\left(\dfrac{5}{y}\right)^2 =$ **17.** $(3a + 2b)(3a - 2b) =$

11. $(6a^3 + 3a^2) \div 3a =$ **18.** $\dfrac{9x^2y^2 - 4a^2b^2}{(3xy - 2ab)} =$

Answers to Practice Problems

8-A
1. T
2. T
3. T
4. T
5. F
6. T
7. F
8. T
9. T
10. T

8-B
1. $2x$
2. $8a$
3. $2a$
4. $7y$
5. $9y$
6. $5a + 3c$
7. $5y$
8. $6V_1$
9. $3R_1 + 4R_3$
10. $7I_1$

8-C
1. $8b^3$
2. $16y^2$
3. $125x^3$
4. a^{15}
5. $9c^2$
6. $16z^4$
7. $9a^2$
8. 1

9. $25a^4$
10. $8x^3$

8-D
1. $2b$
2. $4y$
3. a^3
4. a^2
5. $5y^2$
6. a^7
7. $3.16a$
8. $-2a$
9. $2a^2$
10. y^4

8-E
1. 36
2. 18
3. 32
4. 6
5. 6
6. 32
7. 36
8. 18
9. 12
10. 6

8-F
1. y^2
2. $6y^2$
3. $8a^3$
4. $8a^3$
5. $6a^9$
6. $6x^7$

7. $15b^4$
8. $8I^2$
9. $6ab$
10. $6x^{(a+b)}$

8-G
1. $2a^2$
2. $3a^2$
3. $3.5x^5$
4. $2b^4$
5. $\dfrac{4a}{b}$
6. $2a^{-3}$
7. 1
8. $1.5x^{(a-b)}$

8-H
1. $\dfrac{2x}{5}$
2. $\dfrac{a}{3}$
3. $\dfrac{6a + 1}{4}$
4. $\dfrac{xy}{6}$
5. $\dfrac{2a}{3b}$
6. 1
7. $\dfrac{a^2b}{8}$

8. $\dfrac{x^2y^2}{a^2} = \left(\dfrac{xy}{a}\right)^2$

9. cd

10. $\dfrac{4}{b^2}$

11. $\dfrac{a^3}{b^3}$

12. $\dfrac{a}{b}$

13. $4a^2$

14. $\dfrac{2}{b}$

8-I **1.** $2x$ and 3
2. $5y$ and -4
3. abc and a
4. xy^2 and 8
5. R_1 and R_2
6. C_1 and C_2
7. $ab + xy$
8. $2 + ab$

8-J **1.** 2 and x
2. 5 and y
3. a, b, and c
4. x and y^2
5. R_1 and R_2
6. 2, π, f, and L

7. I and R
8. V and I

8-K **1.** 20
2. 9
3. 80
4. 21
5. 100
6. 29
7. 13
8. 14

8-L **1.** $2a^3b$
2. $2a^3b + 4a^2$
3. $2a^3b - 4a^2$
4. $2ab + 4$
5. $4c$
6. $2y^2 + y$
7. $x^2 + 2xy + y^2$
8. $x^2 - y^2$
9. $x^2 - 2xy + y^2$
10. $2x$

8-M **1.** $4b$
2. b

3. $\dfrac{a + b}{a}$ or

$1 + \dfrac{b}{a}$

4. bd
5. xy
6. $b + 1$
7. ab
8. $3b$
9. $2b$
10. $2a$

8-N **1.** $x^2 - 9$
2. $x^2 + 6x + 9$
3. $x^2 - 6x + 9$
4. $9a^2 - 4b^2$
5. $(x - 2)(x + 2)$
6. $(x + 3)^2$
7. $(x - 3)^2$
8. $(3a + 2b)(3a - 2b)$

8-O **1.** $x^2 + x$
2. $3x^2 + 2x$
3. $a + b$
4. $3x^2 + 9$
5. $x + y$
6. $x + y$
7. $4a + 3$
8. $4a + 3$

9 METHODS OF SOLVING EQUATIONS

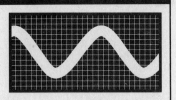

An equation represents a balance of quantities. That is, the combined terms on one side of an equal sign must be equal to the combined terms on the other side of the equal sign. The quantities on either or both sides of the equal sign may contain quantities whose value is not known.

Example $x = 8 - 5$

Answer The unknown quantity to the left of the equal sign is given a letter symbol. (By custom the unknown quantity is given the letter x or y or z, but actually any letter can be used.) We know $8 - 5$ is 3, and since both sides of the equation must be equal, then

$$x = 3$$

What we have just done is *solve the equation;* we have found the value of the unknown quantity x.

More details are explained in the following sections:

9-1 Operations on Both Sides of an Equation
9-2 Transposing Terms
9-3 Solution of Numerical Equations
9-4 Inverting Factors in Literal Equations

9-1 Operations on Both Sides of an Equation

By rearranging the quantities on both sides of the equal sign, we can solve an equation. In general, we can perform almost any arithmetic operation on one side of the equation as long as we perform the same operation on the other side. The only forbidden operation is dividing by zero.

Addition and Subtraction Any positive or negative quantity can be added to both sides of an equation without changing the equality.

Example In the equation $x = 2$, we can add 3 to both x and 2:

Answer $x + 3 = 2 + 3$

The original equation says that x is 2. The next equation says that $x + 3$, or 5, is equal to $2 + 3$, or 5. For both equations, x is 2 and the left side equals the right side. Thus, the addition of 3 to both sides of the equation had no effect on the value of x.

Practice Problems 9-A
Answers at End of Chapter

Add -2 to both sides in the following equations.

1.	$x + 2 = 5$	**4.**	$a = b$
2.	$y + 2 = 9$	**5.**	$1 + 2 = 6$
3.	$x^2 + 2 = 18$	**6.**	$c + 2 = -6$

Reversing the Sides Since the equality applies both ways, the left and right sides of an equation can be interchanged.

Examples $x = 2 + 3$ is the same equality as

$$2 + 3 = x$$

$$x + 5 = 2x + 2 \text{ is the same as}$$

$$2x + 2 = x + 5$$

Multiplication Any number can be used to multiply both sides of an equation. However, every term in the equation must be multiplied by this number. If only some terms are multiplied and others are not, the original equation will be changed and the value of the unknown quantity will be affected.

Example $y = 3$

If we multiply both sides by 5, we get

$$y \times 5 = 3 \times 5$$

Since y is 3 from the original equation, we find 3×5 on the left and 3×5 on the right. Both sides equal 15.
If we had multiplied only the left side of the equation, we would have gotten

$$y \times 5 = 3$$

But since y is 3, the new "equation" would be $3 \times 5 = 3$, which of course is false.
If we had multiplied only the right side of the equation, we would have gotten

$$y = 3 \times 5$$

or $y = 15$

But y was 3 in our original equation, so we have changed the value of y.

Practice Problems 9-B
Answers at End of Chapter

Multiply both sides by 4 in the following equations.

1. $\dfrac{x}{4} = 1$ 4. $\dfrac{x}{4} = 2 + 3$

2. $\dfrac{y}{4} = 2$ 5. $\dfrac{v}{4} = \dfrac{2}{4}$

3. $\dfrac{x}{4} = 0$ 6. $0.25y = 3$

Division Just as we can multiply every quantity in an equation by a number without changing the equation, so too can we divide every quantity by a number (except zero) and leave the equation unchanged.

Example $3x = 6$

Divide both sides of the equation by 3.

Answer Dividing by 3,

$$\frac{3x}{3} = \frac{6}{3}$$

$$x = 2$$

Note that this division actually led to the solution of the equation. By dividing we found the value of x was 2.

Practice Problems 9-C
Answers at End of Chapter

Divide both sides by 4 in the following equations:

1. $4x = 4$ 4. $4I = 9$
2. $4y = 16$ 5. $4x = 8a$
3. $4V = 12$ 6. $4x = 8 + 4$

Reciprocals You can take the reciprocal of both sides of an equation. However, the entire side must be inverted, not just a part.

Example Find the value of x in the following equation:

$$\frac{1}{x} = \frac{1}{2 + 3}$$

Answer If we take the reciprocal of both sides of the equation, we obtain

$$x = 2 + 3$$

or $x = 5$

Note that this is not the same as

$$\frac{1}{x} = \frac{1}{2} + \frac{1}{3}$$

The reason is that the fraction bar is a sign of grouping for the terms $2 + 3$.

Practice Problems 9-D
Answers at End of Chapter

Take the reciprocals of both sides in the following equations.

1. $\dfrac{1}{x} = \dfrac{1}{5}$ 4. $\dfrac{1}{x} = \dfrac{1}{4+5}$

2. $\dfrac{1}{y} = \dfrac{1}{4}$ 5. $\dfrac{1}{a} = \dfrac{1}{10^{-6}}$

3. $\dfrac{1}{x} = \dfrac{1}{2a}$ 6. $\dfrac{1}{x} = 5 \times 10^{5}$

Powers and Roots Finding powers and roots is often useful when solving equations. You can take any power or root provided the same operation is performed on both sides of the equation.

Example	Solve $x^2 = 9 + 7$.
Answer	First combine terms:
	$$x^2 = 16$$
	Take the square root of both sides of the equation:
	$$\sqrt{x^2} = \sqrt{16}$$ $$x = 4$$

Note, however, that $(-4) \times (-4) = 16$, and so -4 is also a solution of the equation. Square roots always have both plus and minus roots. The physical conditions of the problem will usually establish whether the positive or negative root is appropriate or whether they are both valid.

Practice Problems 9-E
Answers at End of Chapter

Take the square root of both sides in the following equations.

1. $x^2 = 16$ 4. $x^2 = (2+3)^2$
2. $y^2 = 9$ 5. $I^2 = 19 + 6$
3. $x^2 = 4b^2$ 6. $a^2 = b^2$

9-2 Transposing Terms

Transposing means moving a term from one side of the equation to the other. When a term is transposed, its sign must be changed, from $+$ to $-$ or from $-$ to $+$.

The transposing of terms is often necessary for the numerical solution of an equation.

Example	Solve the equation
	$$x + 4 = 10$$
Answer	If the $+4$ is transposed to the right side of the equation, its sign is changed and the new equation is
	$$x = 10 - 4$$ $$x = 6$$

The reason why the sign must be changed in transposing terms is that the equality must be maintained. In this example, transposing 4 is the same as adding -4 to both sides. On the left side $+4$ and -4 cancel each other, leaving only the x term.

Practice Problems 9-F
Answers at End of Chapter

Transpose all the literal terms to the left side and all the numbers to the right side in the following equations.

1. $x - 2 = 4$
2. $y^2 - 4 = 5$
3. $3x + 7 = 2x + 8$
4. $2y + 2 = y - 3$
5. $2a - 5 = b + 3$
6. $x + y^2 = y^2 + 7$
7. $V + 3 = 5$
8. $I - 6 = 2$

9-3 Solution of Numerical Equations

The first operation in solving equations is to group all unknowns on one side of the equal sign and all numbers on the other. Then combine terms for one numerical value and one term for the unknown. Finally, divide both sides of the equation by the numerical coefficient of the unknown.

> **Example**   Solve the equation
>
> $$6x - 8 = 4x + 2$$
>
> **Answer**   Transpose the unknown terms to the left side and all numbers to the right side. Remember to change the sign of a transposed term.
>
> $$6x - 4x = 2 + 8$$
>
> Combine terms:
>
> $$2x = 10$$
>
> Divide by 2 (the coefficient of x):
>
> $$\frac{2x}{2} = \frac{10}{2}$$
>
> $$x = 5$$
>
> Check the answer by substituting 5 in the original equation:
>
> $$6x - 8 = 4x + 2$$
> $$6(5) - 8 = 4(5) + 2$$
> $$30 - 8 = 20 + 2$$
> $$22 = 22$$
>
> The fact that each side has the same value shows that the solution of $x = 5$ satisfies the equality.

There are several additional points to keep in mind when solving the next group of equations. When the equation has signs of grouping, they must be removed before combining terms. The reason is that only terms can be transposed, not factors. The terms 0 and 1 can be transposed the same way as other numbers. Finally, a numerical solution can have a negative value.

Practice Problems 9-G
Answers at End of Chapter

Solve the following equations for x.

1. $6x = 4x + 8$
2. $4a - 32 = 2a$
3. $6x - 3 = 2x + 21$
4. $8x - 6 = 12x - 14$
5. $5x - 3 - 12 - 8x = 0$
6. $14 + 5 + 6 = 3x + 2x$
7. $5(x - 2) = 30$
8. $5 - (x - 2) = 30$
9. $4(x - 1) + 8 = 6$
10. $(2x + 4)(2x - 4) = 4x^2 + 2x$

The method of transposing terms can also be used for solving literal equations that have two or more terms on one side. This idea is illustrated in the next set of problems.

> **Example**   Solve the equation $a + b = c - a$ for the value of a.
>
> **Answer**   In this equation all the a terms are transposed to the left side of the equation and the non-a terms to the right side.
>
> $$a + a = c - b$$
> $$2a = c - b$$
>
> Dividing both sides by 2 because it is the coefficient of a,
>
> $$\frac{2a}{2} = \frac{c - b}{2}$$
>
> $$a = \frac{c - b}{2}$$

Practice Problems 9-H
Answers at End of Chapter

Solve the following equations.

1. $c = a + b$ (Solve for a.)
2. $R_T = R_1 + R_2$ (Solve for R_1.)
3. $V_T = V_1 + V_2$ (Solve for V_1.)
4. $R^2 = Z^2 - X^2$ (Solve for Z^2.)
5. $Z^2 = R^2 + X^2$ (Solve for Z.)
6. $C_T = C_1 + C_2$ (Solve for C_2.)

9-4   Inverting Factors in Literal Equations

Many electronics formulas are in the form of a literal equation that has only one term on each side, but the term itself may have factors. An important example is $V = IR$ for Ohm's law. The general form of such a formula can be stated as the equation $a = bx$. In order to solve for x alone, the factor b must be moved to the other side of the equation. However, b here is a factor, not a term. The rules for moving a factor are as follows:

1. The factor becomes inverted between numerator and denominator on opposite sides of the equation.
2. The sign of the inverted factor is *not changed*.

When no denominator is shown, it actually is 1. As an example, equation $a = bx$ can be written with denominators as

$$\frac{a}{1} = \frac{bx}{1}$$

In this example, the factor b in the numerator at the right becomes inverted as a factor of 1 in the denominator at the left. The solution for x then is $a/b = x$, or $x = a/b$. This operation is equivalent to dividing both sides by b in the original equation. For this reason, the sign is not changed when inverting a factor from one side to the other, as there is no transposing of terms.

The rule for inverting factors means that the numerator on one side of the equation and the denominator on the other side can be cross-multiplied.

Example $\dfrac{a}{b} = \dfrac{c}{d}$

Show this equation in another form with two factors on each side of the equation. Then solve for a, b, c, and d individually.

Answer   The equation above when cross-multiplied becomes

$$ad = cb$$

Solving for a,

$$a = \frac{cb}{d}$$

Solving for d,

$$d = \frac{cb}{a}$$

Solving for b,

$$\frac{ad}{c} = b \qquad \text{or} \qquad b = \frac{ad}{c}$$

Solving for c,

$$\frac{ad}{b} = c \qquad \text{or} \qquad c = \frac{ad}{b}$$

In summary, a factor can be moved *on the diagonal,* from numerator on one side to denominator on the other side, or the opposite way, without a change in sign.

Practice Problems 9-I
Answers at End of Chapter

Solve the following equations.

1. $ab = c$ (Solve for a.)
2. $a + b = c$ (Solve for a.)
3. $V = IR$ (Solve for I.)
4. $I = \dfrac{V}{R}$ (Solve for V.)
5. $P = VI$ (Solve for I.)
6. $I = \dfrac{P}{V}$ (Solve for P.)
7. $Q = CV$ (Solve for C.)
8. $C = \dfrac{Q}{V}$ (Solve for Q.)
9. $P = I^2R$ (Solve for I.)
10. $l = \pi d$ (Solve for d.)
11. $y = abx$ (Solve for x.)
12. $X = 2\pi fL$ (Solve for L.)
13. $y = \dfrac{a - b}{x}$ (Solve for x.)
14. $4x^2 = 100$ (Solve for x.)

Review Problems
Answers to Odd-Numbered Problems at Back of Book

The following problems summarize the methods of solving equations. Solve for x.

1. $3x + 4 = x + 10$

2. $x^2 + 3 = 19$

3. $\dfrac{x}{3} + 2 = 7$

4. $\dfrac{1}{x} = 3 + 4$

5. $x^3 = (1 + 8)^3$

6. $2(x + 5) = 18$

7. $\dfrac{x^2 - 9}{x + 3} = 5$

8. $x = \dfrac{a^2 bcd}{abcd}$

9. $\dfrac{4a^2}{9b^2} = x^2$

10. $y = abx$

Answers to Practice Problems

9-A
1. $x = 3$
2. $y = 7$
3. $x^2 = 16$
4. $a - 2 = b - 2$
5. $I = 4$
6. $c = -8$

9-B
1. $x = 4$
2. $y = 8$
3. $x = 0$
4. $x = 20$
5. $v = 2$
6. $y = 12$

9-C
1. $x = 1$
2. $y = 4$
3. $V = 3$
4. $I = 2.25$
5. $x = 2a$
6. $x = 3$

9-D
1. $x = 5$
2. $y = 4$
3. $x = 2a$
4. $x = 9$
5. $a = 10^6$
6. $x = 0.2 \times 10^{-5}$

9-E
1. $x = 4$
2. $y = 3$
3. $x = 2b$
4. $x = 5$

5. $I = 5$
6. $a = b$

9-F
1. $x = 6$
2. $y^2 = 9$
3. $x = 1$
4. $y = -5$
5. $2a - b = 8$
6. $x = 7$
7. $V = 2$
8. $I = 8$

9-G
1. $x = 4$
2. $a = 16$
3. $x = 6$
4. $x = 2$
5. $x = -5$
6. $5 = x$ or $x = 5$
7. $x = 8$
8. $x = -23$
9. $x = 0.5$
10. $-8 = x$ or $x = -8$

9-H
1. $a = c - b$
2. $R_1 = R_T - R_2$
3. $V_1 = V_T - V_2$
4. $Z^2 = R^2 + X^2$
5. $Z = \sqrt{R^2 + X^2}$
6. $C_2 = C_T - C_1$

9-I
1. $a = \dfrac{c}{b}$

2. $a = c - b$
3. $I = \dfrac{V}{R}$
4. $V = IR$
5. $I = \dfrac{P}{V}$
6. $P = IV$
7. $C = \dfrac{Q}{V}$
8. $Q = CV$
9. $I = \sqrt{\dfrac{P}{R}}$
10. $d = \dfrac{l}{\pi}$
11. $x = \dfrac{y}{ab}$
12. $L = \dfrac{X}{2\pi f}$
13. $x = \dfrac{a - b}{y}$
14. $x = 5$

10 SIMULTANEOUS LINEAR EQUATIONS

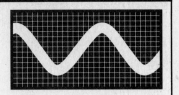

A linear equation has the unknown only in the first power, without any higher exponents. An example is $x + 3 = 5$. The linear description means that a graph of the equation is a straight line.

Simultaneous linear equations form a group or set with the same unknowns. The number of equations must equal the number of unknowns to determine the solution. The following example shows a set of two linear equations for the two unknowns x and y:

$$x + y = 8$$
$$x - y = 2$$

These are simultaneous equations because the solution for the unknown quantities results in the same values for both equations. The solution here is $x = 5$ and $y = 3$. Both unknowns can be determined because there are two statements in the equations that specify the values for x and y.

In electronics, simultaneous linear equations are used to solve problems in the application of Kirchhoff's* laws for the voltages and currents in a circuit. The method is to set up equations for the voltages around different paths and solve for the currents and voltages.

Several methods for the solution of simultaneous linear equations are explained here. They all result in the same answer. The method to use is only a question of which seems more convenient.

More details are explained in the following sections:

10-1   Adding or Subtracting Equations
10-2   Method of Substitution for One Unknown
10-3   Determinants for Two Equations
10-4   Determinants for Three Equations

10-1 Adding or Subtracting Equations

Consider the example given before for a set of two simultaneous linear equations:

*See Bernard Grob, *Basic Electronics*, Chap. 9, Macmillan/McGraw-Hill School Publishing Company.

$$x + y = 8$$
$$x - y = 2$$

The two equations can be added, or one can be subtracted from the other. The result will still have the left side equal to the right side because the equality has not been changed. For addition, just add all the similar terms. Then

$$x + y = 8$$
$$\underline{x - y = 2}$$
$$2x \pm 0 = 10$$

The y terms drop out to allow an equation in x only.

$$2x = 10$$
$$\frac{2x}{2} = \frac{10}{2}$$
$$x = 5$$

Then, to find y, just substitute the value $x = 5$ in either of the two original equations. The result for the first equation in the set is $y = 8 - 5 = 3$.

If the original two equations are subtracted, the x term will drop out. Then

$$x + y = 8$$
$$\underline{x - y = 2}$$
$$0 + 2y = 6$$
$$2y = 6$$
$$\frac{2y}{2} = \frac{6}{2}$$
$$y = 3$$

To subtract, change all the signs in the equation used as the subtrahend and then add similar terms. With y known to be 3, we can solve for x in the first original equation as $x = 8 - 3 = 5$.

In some cases, it may be necessary to change the coefficients to get equal numerical values that can drop out. Remember that an equation can be multiplied or divided by any number except zero without changing the equality. However, this must be done for all terms in the equation.

Example   Solve for I_1 and I_2.

$$3I_1 + I_2 = 14$$
$$2I_1 + 3I_2 = 7$$

Multiply the first equation by 3. Then the set is

$$9I_1 + 3I_2 = 42$$
$$2I_1 + 3I_2 = 7$$

Now subtract the second equation:

$$9I_1 + 3I_2 = 42$$
$$2I_1 + 3I_2 = 7$$
$$\overline{7I_1 \pm 0 = 35}$$

$$7I_1 = 35$$

$$I_1 = \frac{35}{7} = 5$$

To find I_2, finally, substitute 5 for I_1 in the first original equation:

$$3(5) + I_2 = 14$$
$$I_2 = 14 - 15$$
$$I_2 = -1$$

Answer $I_1 = 5$
        $I_2 = -1$

Practice Problems 10-A
Answers at End of Chapter

Solve the following pairs of equations by the method of adding or subtracting equations.

1. $2x - 2y = 4$
     $2x + 2y = 12$
2. $3x + 3y = 24$
     $3x - 3y = 6$
3. $3x + y = 14$
     $2x + 3y = 7$

4. $3x + y = 1$
     $5x - 2y = 9$

10-2   Method of Substitution for One Unknown

Consider the original set of two equations for an example,

$$x + y = 8$$
$$x - y = 2$$

From the second equation, x can be stated as $2 + y$, by transposing. Use this value for x in the first equation. Then

$$2 + y + y = 8$$
$$2y = 8 - 2 = 6$$
$$y = 3$$

Substitute 3 for y in either of the original equations to find $x = 5$. These calculations are

$$x + 3 = 8 \quad \text{or} \quad x = 8 - 3$$
$$\text{so that} \quad x = 5$$

In the other equation,

$$x - 3 = 2 \quad \text{or} \quad x = 2 + 3$$
$$\text{so that} \quad x = 5$$

Practice Problems 10-B
Answers at End of Chapter

Solve the equations in Practice Problems 10-A by the method of substitution.

10-3   Determinants for Two Equations

The method of determinants is probably the best way of solving simultaneous linear equations because it uses the numerical values in the equations to give the numerical solution directly. Consider this example of two equations that was solved before:

$$3I_1 + I_2 = 14$$
$$2I_1 + 3I_2 = 7$$

First, we use the numerical coefficients of the unknowns. The constant values 14 and 7 are not used yet. Write the coefficients in this form:

$$\begin{vmatrix} 3 & 1 \\ 2 & 3 \end{vmatrix}$$

The vertical lines show that the numbers are in a group or set called a *matrix*. The matrix can be defined mathematically as a rectangular arrangement of rows and columns for numbers. The top horizontal row in the matrix is for the first original equation, and the bottom row is for the second equation. Each vertical column has the numerical coefficients for like terms in both equations.

Next, we expand the matrix to determine its numerical value. This is done by cross multiplying values on the diagonal, as shown here.

Add this cross product $\begin{vmatrix} 3 & 1 \\ 2 & 3 \end{vmatrix}$ $3 \times 3 = 9$

Subtract this cross product $\begin{vmatrix} 3 & 1 \\ 2 & 3 \end{vmatrix}$ $2 \times 1 = 2$

$$9 - 2 = 7$$

The 9 is the result of multiplying 3×3 on the diagonal down to the right. The 2 is 2×1 from multiplying on the diagonal up to the right. This product is subtracted, with the result $9 - 2 = 7$. The 7 is the *determinant* of the system, meaning that it is used to set up the determinants for the unknowns I_1 and I_2.

The next step is to calculate I_1 and I_2 from their determinants. In general form, I_1 and I_2 can be stated as

$$I_1 = \frac{\text{determinant}_1}{\text{system determinant}} = \frac{\text{determinant}_1}{7}$$

$$I_2 = \frac{\text{determinant}_2}{\text{system determinant}} = \frac{\text{determinant}_2}{7}$$

The 7 in the denominator is the value of the determinant for this system of equations, which has already been calculated. This value is used in the calculations

for both I_1 and I_2. In the numerator, the specific matrix of determinants for I_1 is

$$\text{Determinant}_1 = \begin{vmatrix} 14 & 1 \\ 7 & 3 \end{vmatrix}$$

In this determinant, the numerical coefficients for I_1, 3 and 2, are replaced by the constant values in the original equations, 14 and 7. The method of expanding this matrix is the same as before, using the difference between the diagonal cross products. Then

$$\text{Determinant}_1 = \begin{vmatrix} 14 & 1 \\ 7 & 3 \end{vmatrix} = (14 \times 3) - (7 \times 1)$$

$$= 42 - 7 = 35$$

$$I_1 = \frac{35}{7} = 5$$

For I_2, replace its numerical coefficients by the constant values in the original equation. Then

$$\text{Determinant}_2 = \begin{vmatrix} 3 & 14 \\ 2 & 7 \end{vmatrix} = (3 \times 7) - (2 \times 14)$$

$$= 21 - 28 = -7$$

$$I_2 = \frac{-7}{7} = -1$$

The complete solution, therefore, is $I_1 = 5$ and $I_2 = -1$, the same as the solution found before by subtraction of the equations. Actually, the formula for determinants is derived from subtraction of the equations in their general form.

The explanation may make the method of determinants seem longer than other methods of solution, but the work is actually shorter. Let us solve another pair of equations that we solved before by the method of substitution.

Example   Solve for x and y.

$$x - y = 2$$
$$x + y = 8$$

Answer   By the method of determinants, the solution is

$$x = \frac{\begin{vmatrix} 2 & 1 \\ 8 & 1 \end{vmatrix}}{\begin{vmatrix} 1 & -1 \\ 1 & 1 \end{vmatrix}} = \frac{[2 \times 1] - [8 \times (-1)]}{[1 \times 1] - [1 \times (-1)]}$$

$$= \frac{2 + 8}{1 + 1} = \frac{10}{2} = 5$$

$$y = \frac{\begin{vmatrix} 1 & 2 \\ 1 & 8 \end{vmatrix}}{2} = \frac{(1 \times 8) - (1 \times 2)}{2}$$

$$= \frac{8 - 2}{2} = \frac{6}{2} = 3$$

Note that $-y$ in the first equation has the numerical coefficient of -1. Also, subtracting a negative cross product is the same as adding a positive value. The final answers are $x = 5$ and $y = 3$, the same as those found by the method of substitution.

Practice Problems 10-C
Answers at End of Chapter

Solve the equations in Practice Problems 10-A by the method of determinants.

10-4 Determinants for Three Equations

The method is similar to that for solving two equations, but the determinants have three rows and columns. Let us solve the following set of equations:

$$\begin{array}{rrrr} 15I_A - & 5I_B & & = 16 \\ -5I_A + & 20I_B - & 5I_C = & 0 \\ & -5I_B + & 15I_C = & 4 \end{array}$$

Using three sets of numerical coefficients, the determinant for the system is

$$\begin{vmatrix} 15 & -5 & 0 \\ -5 & 20 & -5 \\ 0 & -5 & 15 \end{vmatrix}$$

Note that a zero is used where no term exists.

To expand this determinant, there is a question about cross multiplying on the diagonal, since only the middle diagonal includes three coefficients. A useful technique is to rewrite the determinant with the first two columns repeated after the third column, like this:

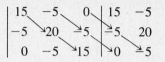

Start with the first number in the upper left and follow the arrows down to the right for three cross products of three coefficients each. These cross products are to be added:

$$[(15 \times 20 \times 15)] + [(-5) \times (-5) \times 0]$$
$$+ [0 \times (-5) \times (-5)] = 4500 + 0 + 0 = 4500$$

For the diagonal cross products to be subtracted, follow the dotted arrows up and to the right, starting with the coefficient at the bottom left. These cross products are added and the sum is subtracted from the previous cross-product sum. The calculations for the cross products combined for subtractions are:

$$\begin{vmatrix} 15 & -5 & 0 \\ -5 & 20 & -5 \\ 0 & -5 & 15 \end{vmatrix} \begin{matrix} 15 & -5 \\ -5 & 20 \\ 0 & -5 \end{matrix}$$

The result is

$$[0 \times 20 \times 0] + [(-5) \times (-5) \times 15]$$
$$+ [15 \times (-5) \times (-5)]$$
$$= 0 + 375 + 375$$
$$= 750$$

The determinant of this system, then, is

$$4500 - 750 = 3750$$

Use this value of 3750 in the denominator of the determinants for the unknowns.

Now we can solve for the unknowns I_A, I_B, and I_C. In each case, the determinant for the unknown uses the numerical constants of the original equations in place of its numerical coefficients.

$$I_A = \dfrac{\begin{vmatrix} 16 & -5 & 0 \\ 0 & 20 & -5 \\ 4 & -5 & 15 \end{vmatrix}}{3750}$$

$$= \dfrac{4900 - 400}{3750} = \dfrac{4500}{3750}$$

$$I_A = 1.2$$

To solve for the next unknown, I_B, substitute the constant values in the equations for the numerical coefficient of I_B. This determinant is

$$I_B = \dfrac{\begin{vmatrix} 15 & 16 & 0 \\ -5 & 0 & -5 \\ 0 & 4 & 15 \end{vmatrix}}{3750}$$

$$= \dfrac{0 - (-1500)}{3750} = \dfrac{1500}{3750}$$

$$I_B = 0.4$$

Finally, the determinant for I_C is

$$I_C = \dfrac{\begin{vmatrix} 15 & -5 & 16 \\ -5 & 20 & 0 \\ 0 & -5 & 4 \end{vmatrix}}{3750}$$

$$= \dfrac{1600 - 100}{3750} = \dfrac{1500}{3750}$$

$$I_C = 0.4$$

As a check on this solution, substitute the values of 1.2 for I_A and 0.4 for I_B and I_C in the original set of equations. Use the middle equation because three unknowns.

$$-5I_A + 20I_B - 5I_C = 0$$
$$-5(1.2) + 20(0.4) - 5(0.4) = 0$$
$$-6 + 8 - 2 = 0$$
$$0 = 0$$

The equality shows that the values are correct.

Practice Problems 10-D
Answers at End of Chapter

Solve the following sets of three equations by the method of determinants.

1. $x + y + z = 6$
$5x + 4y + 3z = 22$
$3x + 4y - 3z = 2$

2. $x + 3y + 6z = 32$
$x + 2y - 3z = 81$
$x - 7y + 4z = -94$

Review Problems
Answers to Odd-Numbered Problems at Back of Book

Solve the following sets of equations for the unknown values. Any method can be used.

1. $x - y = 2$
$x + y = 6$

2. $2a - 2b = 0$
$2a + 2b = 10$

3. $5I_1 + I_2 = 7$
$7I_1 + 3I_2 = 9$

4. $x - y = 8$
$x + y = 12$

5. $3V_1 + V_2 = 1$
$5V_1 - 2V_2 = 9$

6. $2x - 5y = 1$
$3x + 2y = 11$

7. $4x + y = 30$
$5x + y = 38$

8. $6I_A - 2I_B + 0 = 12$
$-2I_A + 8I_B - 2I_C = 0$
$0 - 2I_B + 6I_C = -8$

Answers to Practice Problems

10-A
1. $x = 4$
$y = 2$
2. $x = 5$
$y = 3$
3. $x = 5$
$y = -1$
4. $x = 1$
$y = -2$

10-B   See Answers to Practice Problems 10-A

10-C   See Answers to Practice Problems 10-A

10-D
1. $x = 1$
$y = 2$
$z = 3$
2. $x = 32$
$y = 14$
$z = -7$

11 TRIGONOMETRY

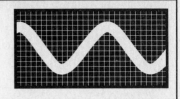

Trigonometry is the study of triangles and angles. A triangle is made up of three straight lines, called the *sides* of the triangle. Where two sides meet, they enclose an angle. The three sides enclose three angles.

More generally, angles are needed as units to specify the amount of rotational motion. Just as we say an object is moved 8 ft or 8 in along a linear path, for example, the object can move 90° for rotational motion in a circular path. The 90° is one-quarter of a full circle.

More details are explained in the following sections:

11-1 Types of Angles

An angle is formed when two lines meet. In Fig. 11-1, when the line OP_1 is hinged at the origin O and rotates to the position P_2, the line generates the angle P_1OP_2. The Greek letter symbols θ (theta) and ϕ (phi) are generally used with the angle sign $\angle$. In this example P_1OP_2 is indicated as $\theta = 60°$.

The unit of 1°, or one degree, is $1/360$ of the rotation around a circle. The complete circle, therefore, consists of 360°. In this example, θ is 60°, because the amount of rotation is $1/6$ of a complete circle, equal to 360°/6, or 60°.

By convention, counterclockwise rotation, as in Fig. 11-1, is the positive direction for measuring angles. This rotation corresponds to motion upward or to the right for the positive direction for linear measurements.

Example How many degrees are there in $1/8$ rotation of a circle?

Answer Since the complete circle contains 360°,

$$\frac{1}{8} \times 360° = 45°$$

A negative angle can be generated by clockwise rotation. In Fig. 11-1, if the line OP_2 were rotated the same amount but in the opposite direction, the angle would be $-60°$.

Practice Problems 11-A
Answers at End of Chapter

How many degrees are there in each of the following rotations?

1.	¼ circle	**5.**	⅓ circle
2.	½ circle	**6.**	⅙ circle
3.	¾ circle	**7.**	⅕ circle
4.	¹⁄₁₀ circle	**8.**	Full circle

The angle of 90° is called a *right angle*. One side is upright compared to the other, as shown in Fig. 11-2. Since 90° is one-quarter of a circle, the perpendicular sides are in *quadrature*.

Angles of less than 90° are *acute*. For instance, 60° is

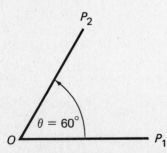

Fig. 11-1 Generating the positive angle θ by counterclockwise rotation from P_1 to P_2.

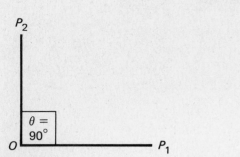

Fig. 11-2 A 90° angle is a right angle with perpendicular sides, in quadrature.

an acute angle. An angle of more than 90° is *obtuse*. An example is 120°.

Practice Problems 11-B
Answers at End of Chapter

State whether each of the following is an acute, obtuse, or right angle.

1.	17°	**3.**	90°	**5.**	45°	**7.**	1°
2.	64°	**4.**	150°	**6.**	100°	**8.**	89°

Two angles whose sum is equal to 90° are *complementary*. For instance, 30° and 60° are complementary angles; one angle is the complement of the other. Two angles whose sum is equal to 180° are *supplementary*. Examples of supplementary angles are 120° and 60°.

Practice Problems 11-C
Answers at End of Chapter

Give the complement of the following angles.

1.	30°	**3.**	45°	**5.**	20°	**7.**	40°
2.	60°	**4.**	17°	**6.**	53°	**8.**	50°

An angle of less than 1° can be divided into decimal parts. For instance, 0.5° is one-half a degree. Also, 26.5° is midway between 26° and 27°. This method of describing an angle in decimal fractions is called *decitrig* notation.

Angles may also be divided into minutes and seconds. There are 60 minutes in a degree and 60 seconds in a minute. This method of dividing an angle into 60 subdivisions is called *sexagesimal* notation.

Example   Write 26°30′0″ using decitrig notation.

Answer   Since $30' = \frac{1}{2}° = 0.5°$,

$$26°30'0'' = 26.5°$$

Another unit for measuring angles is the *grad,* equal to $\frac{1}{400}$ of a circle. Therefore, 1 grad = $\frac{360}{400}$ = 0.9°.

The decitrig form of notation, such as 26.5°, is used with electronic calculators. Make sure the DRG key on the calculator is set on the D position for decitrig. Calculations with angles are much simpler in decitrig form.

To add or subtract angles in decitrig form, just combine their numerical values.

Example   Add 26.5° and 32.7°.

Answer   Align the decimal points and add as usual.

$$\begin{array}{r} 26.5° \\ +32.7° \\ \hline 59.2° \end{array}$$

Practice Problems 11-D
Answers at End of Chapter

Add or subtract the following angles, as indicated.

1.	30° + 10° =	**6.**	120° − 70° =
2.	12.5° + 8° =	**7.**	2° + 3° =
3.	84° − 14° =	**8.**	24° − 30° =
4.	14.5° + 2°30′ =	**9.**	90° − 30° =
5.	70° + 50° =	**10.**	90° − 60° =

11-2 The Right Triangle

In the analysis of ac circuits, the calculations make extensive use of trigonometry with right triangles. The reason is the perpendicular sides of the triangle can represent currents or voltages 90° out of phase. (See Fig. 11-3.) The altitude *a* is perpendicular to the base *b*. These two sides form a right angle. The side opposite the 90° angle is called the *hypotenuse,* indicated by the letter *c*.

The pythagorean theorem from ancient Greek geometry states that the square of the hypotenuse is equal to the sum of the squares of the other two sides. This

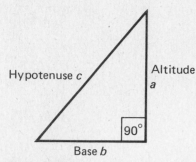

Fig. 11-3 The sides of a right triangle. Hypotenuse c is opposite the 90° angle.

equality is illustrated in Fig. 11-4, with squares for all the sides of the triangle. As a formula,

$$c^2 = a^2 + b^2 \qquad (11\text{-}1)$$

In order to find the length of the hypotenuse, it is necessary to find the square root of c^2. Taking the square roots of both sides of the equation, we get

$$c = \sqrt{a^2 + b^2} \qquad (11\text{-}2)$$

To summarize, the length of the hypotenuse is calculated as follows:

1. Square each of the other two sides.
2. Add the squares.
3. Take the square root of the sum of the squares.

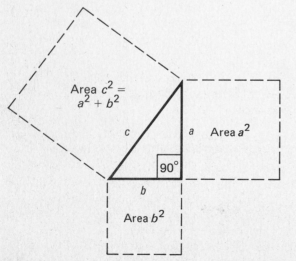

Fig. 11-4 Illustrating the pythagorean theorem for a right triangle: $c^2 = a^2 + b^2$.

Example   In Fig. 11-5a, one side of the right triangle is 4 units long while the other side is 3 units long. Find the length of the hypotenuse.

Answer   Using the formula

$$c = \sqrt{a^2 + b^2}$$

square each side:

$$4^2 = 16$$
$$3^2 = 9$$
$$c = \sqrt{16 + 9} = \sqrt{25}$$

Find the square root of the sum of the squares:

$$c = \sqrt{25} = 5 \text{ units}$$

This type of triangle is called a 3:4:5 right triangle. When the perpendicular sides are in the ratio 3:4, as in Fig. 11-5a, or 4:3, as in Fig. 11-5b, the hypotenuse must be in the ratio of 5.

Example   Assume that two sides of a right triangle are 15 and 20. Find the hypotenuse.

Answer $c = \sqrt{a^2 + b^2} = \sqrt{15^2 + 20^2}$
$c = \sqrt{225 + 400} = \sqrt{625}$
$c = 25$

The three sides of the triangle are thus 15, 20, and 25. If we divided each of these sides by 5,

$$\frac{15}{5} = 3 \qquad \frac{20}{5} = 4 \qquad \frac{25}{5} = 5$$

we would end up with the basic 3:4:5 right triangle. In other words, the actual lengths may vary, but their ratio will remain the same in this type of triangle.

Another special type of right triangle is the triangle with two equal sides shown in Fig. 11-5c. Here the

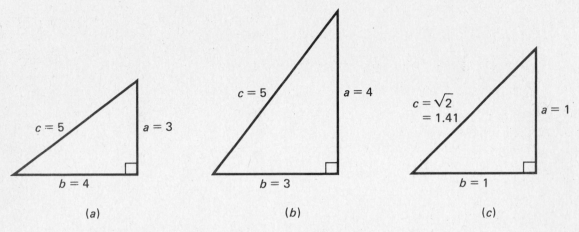

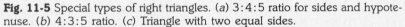

Fig. 11-5 Special types of right triangles. (*a*) 3:4:5 ratio for sides and hypotenuse. (*b*) 4:3:5 ratio. (*c*) Triangle with two equal sides.

hypotenuse must be $\sqrt{2}$, or 1.41, times greater than either side.

Example   Two sides of a right triangle are equal to 10. Find the hypotenuse.

Answer   The formula $c = \sqrt{a^2 + b^2}$ will, of course, lead to the answer. However, the hypotenuse must be 1.41 times the length of one side. Thus

$$c = 1.41 \times 10 = 14.1$$

As a check the formula will be used:

$$c = \sqrt{10^2 + 10^2} = \sqrt{100 + 100}$$
$$c = \sqrt{200} = \sqrt{2 \times 100}$$
$$= \sqrt{2} \times \sqrt{100}$$
$$c = 1.41 \times 10 = 14.1$$

This checks with the original calculation.

Formulas 11-1 and 11-2 apply to any right triangle. As a check on the answer to a right-triangle problem, it should be noted that the hypotenuse must be longer than either of the sides but less than their sum. For instance, in a 3:4:5 triangle the hypotenuse 5 is greater than either side, 3 or 4, but less than their sum of 7. Also, in Fig. 11-5*c*, 1.41 is more than 1 but still less than $1 + 1 = 2$.

Practice Problems 11-E
Answers at End of Chapter

Find the hypotenuse *c* of the following right triangles.

1.	$a = 6, b = 8$	**6.**	$a = 10, b = 1$
2.	$a = 8, b = 6$	**7.**	$a = 1, b = 10$
3.	$a = 5, b = 5$	**8.**	$a = 10, b = 10$
4.	$a = 2, b = 4$	**9.**	$a = 4, b = 5$
5.	$a = 4, b = 2$	**10.**	$a = 2, b = 8$

Formulas 11-1 and 11-2 can be transposed to find one side when the other side and the hypotenuse are given. Then,

$$a = \sqrt{c^2 - b^2} \qquad (11\text{-}3)$$

$$b = \sqrt{c^2 - a^2} \qquad (11\text{-}4)$$

Note that in the radical, a^2 or b^2 must be subtracted from c^2, because the hypotenuse *c* is the longest side of the triangle.

Example   Find the length of the other side of a right triangle in which the hypotenuse is 13 and one side is 12.

Answer $a = \sqrt{c^2 - b^2}$
$a = \sqrt{13^2 - 12^2}$
$= \sqrt{169 - 144}$
$= \sqrt{25}$
$a = 5$

Practice Problems 11-F
Answers at End of Chapter

Find the missing side a or b in the following triangles.

1. $c = 14.14,$
   $a = 10$
2. $c = 14.14,$
   $b = 10$
3. $c = 5,$
   $b = 4$
4. $c = 5,$
   $a = 3$
5. $c = 4.47,$
   $a = 2$
6. $c = 4.47,$
   $b = 2$
7. $c = 6.4,$
   $b = 5$
8. $c = 8.25,$
   $a = 2$

The way in which the right triangle is used for ac analysis is illustrated in Fig. 11-6 for a series circuit. The perpendicular sides represent reactance X and resistance R which are 90° out of phase. Then the hypotenuse is their phasor sum, which is the resultant impedance Z. All these quantities are in units of ohms (Ω).

In Fig. 11-6a, the perpendicular vector is the inductive reactance X_L of a coil. This right triangle is shown upward with an angle of 90°, for the phase between X_L and R. In Fig. 11-6b, the triangle for X_C is shown downward with an angle of −90°, because an X_C voltage lags an R voltage. The X_C is the reactance of a capacitor.

Actually both triangles are diagrams for the addition of phasors, with the tail of one phasor started from the arrowhead of another phasor, with the specified angle. The hypotenuse Z can be found by the pythagorean theorem:

$$Z = \sqrt{R^2 + X^2} \tag{11-5}$$

This formula applies for either X_L or X_C, because the square of a negative value for X_C becomes positive. The only difference is that angle θ is positive for X_L but negative for X_C.

Practice Problems 11-G
Answers at End of Chapter

Find the magnitude of Z, without the phase angle, for the following values of X and R.

1. $X_L = 6, R = 8$
2. $X_C = 6, R = 8$
3. $X_L = 5, R = 5$
4. $X_C = 5, R = 5$
5. $X_L = 2, R = 3.75$
6. $X_C = 4, R = 5$
7. $X_L = 1, R = 10$
8. $X_L = 10, R = 1$
9. $X_L = 4, R = 3$
10. $X_L = 3, R = 4$

An important characteristic of any triangle is that the sum of the three angles must equal 180°. For a right triangle, one angle is 90°. Therefore, the other two angles must total 90°, which makes them complementary. As an example, in Fig. 11-7 with $\theta = 30°$, the complementary angle ϕ is 60°.

It should be noted that θ is in the *standard reference position*, with one side horizontal. Then the opposite side is the altitude a. This horizontal position is assumed to mean angle θ, rather than ϕ, unless specifically indicated otherwise.

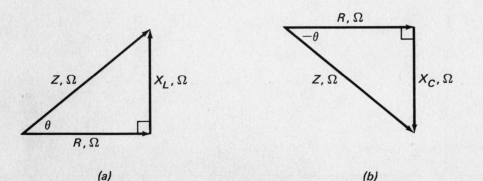

(a)   (b)

Fig. 11-6 Using the hypotenuse of a right triangle for electrical impedance Z. Angle θ is the phase angle of the ac circuit. (a) Sides represent phasors for inductive reactance X_L at 90° with respect to resistance R. (b) Capacitive reactance X_C at −90°.

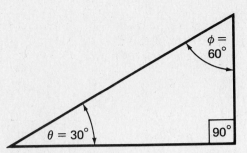

Fig. 11-7 Right triangle showing complementary angles θ and ϕ of 30° and 60°, respectively. They total 90°. Note that θ is the angle in the standard reference position, with one side horizontal.

Practice Problems 11-H
Answers at End of Chapter

Find either angle θ or ϕ, as indicated in the following, for a right triangle.

1.	$\theta = 30°$, $\phi = ?$	**5.**	$\theta = 5°$, $\phi = ?$
2.	$\theta = 60°$, $\phi = ?$	**6.**	$\phi = 5°$, $\theta = ?$
3.	$\phi = 45°$, $\theta = ?$	**7.**	$\theta = 53°$, $\phi = ?$
4.	$\phi = 20°$, $\theta = ?$	**8.**	$\phi = 12.5°$, $\theta = ?$

11-3　Trigonometric Functions

A mathematical function is a fixed relation between two quantities. A trigonometric function defines an angle in a triangle in terms of the sides. Referring to Fig. 11-8, a triangle having sides a, b, and hypotenuse c, for angle θ in the standard reference position, the three main trigonometric functions are defined as follows:

Sine of θ, or sin θ,

$$= \frac{\text{opposite side}}{\text{hypotenuse}} = \frac{a}{c} \qquad (11\text{-}6)$$

Cosine of θ, or cos θ,

$$= \frac{\text{adjacent side}}{\text{hypotenuse}} = \frac{b}{c} \qquad (11\text{-}7)$$

Tangent of θ, or tan θ,

$$= \frac{\text{opposite side}}{\text{adjacent side}} = \frac{a}{b} \qquad (11\text{-}8)$$

Side b is the adjacent side for angle θ because b is one side of the angle. Side a is the opposite side because it is not part of the angle. The hypotenuse is always opposite the right angle.

These formulas can be applied to the 3:4:5 triangle in Fig. 11-8. For the following trigonometric functions of angle θ,

$$\sin \theta = \frac{\text{opposite side}}{\text{hypotenuse}} = \frac{4}{5} = 0.8$$

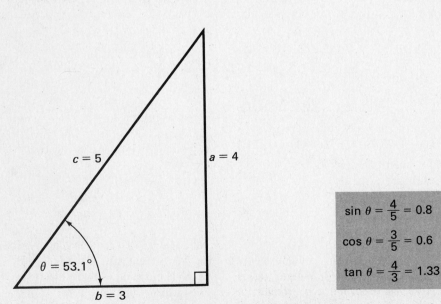

$$\sin \theta = \frac{4}{5} = 0.8$$
$$\cos \theta = \frac{3}{5} = 0.6$$
$$\tan \theta = \frac{4}{3} = 1.33$$

Fig. 11-8 Right triangle for defining the trigonometric functions sin θ, cos θ, and tan θ.

$$\cos \theta = \frac{\text{adjacent side}}{\text{hypotenuse}} = \frac{3}{5} = 0.6$$

$$\tan \theta = \frac{\text{opposite side}}{\text{adjacent side}} = \frac{4}{3} = 1.33$$

Note that the functions are numerical values for the ratio of two sides, not degrees like angle θ. However, the value of the function specifies the angle. This example is for $\theta = 53.1°$, as shown in Fig. 11-8.

The angle of $53.1°$, as one example, has these values for sine, cosine, and tangent in any triangle. Any angle can be specified either in degrees or in terms of its trigonometric functions.

Only one function is needed to specify the angle. The $\sin \theta$, $\cos \theta$, or $\tan \theta$ can be used. For instance, with $\sin \theta = 0.8$ in Fig. 11-8, the angle θ is specified as $53.1°$. Or, the angle for $\cos \theta = 0.6$ is the same $53.1°$. Also, the value of $\tan \theta = 1.33$ means the angle is $53.1°$. Note that $\sin \theta$ or $\cos \theta$ cannot be more than 1 because a or b cannot be more than c.

Practice Problems 11-I
Answers at End of Chapter

Find $\sin \theta$, $\cos \theta$, or $\tan \theta$, as indicated, in the following right triangles with the given opposite side a, adjacent side b, and hypotenuse c:

1. $a = 2$, $c = 2$; $\sin \theta = ?$
2. $b = 1$, $c = 2$; $\cos \theta = ?$
3. $a = 4$, $c = 5$; $\sin \theta = ?$
4. $b = 4$, $c = 5$; $\cos \theta = ?$
5. $a = 6$, $b = 8$, $c = 10$; $\sin \theta = ?$
6. $a = 6$, $b = 8$, $c = 10$; $\tan \theta = ?$
7. $a = 2$, $b = 2$, $c = 2.8$; $\tan \theta = ?$
8. $a = 10$, $b = 10$, $c = 14.14$; $\tan \theta = ?$
9. $a = 5$, $b = 5$, $c = 7$; $\tan \theta = ?$
10. $a = 5$, $b = 5$, $c = 7$; $\sin \theta = ?$

Note that there are three more trigonometric functions, the reciprocals of $\sin \theta$, $\cos \theta$, and $\tan \theta$. These reciprocal functions are

Cotangent of $\theta = \cot \theta$

$$= \frac{\text{adjacent side}}{\text{opposite side}} = \frac{b}{a}$$

$$= \frac{1}{\tan \theta}$$

Secant of $\theta = \sec \theta$

$$= \frac{\text{hypotenuse}}{\text{adjacent side}} = \frac{c}{b}$$

$$= \frac{1}{\cos \theta}$$

Cosecant of $\theta = \csc \theta$

$$= \frac{\text{hypotenuse}}{\text{opposite side}} = \frac{c}{a}$$

$$= \frac{1}{\sin \theta}$$

Reciprocal functions invert the ratio of the two sides. For the tangent and cotangent, as an example, if $\tan \theta = \frac{3}{4}$, then $\cot \theta = \frac{4}{3}$. Or, if $\tan \theta = 0.75$, then $\cot \theta = \frac{1}{0.75} = 1.33$.

Practice Problems 11-J
Answers at End of Chapter

Find the reciprocal functions.

1. $\tan \theta = 2$, $\cot \theta = ?$
2. $\cot \theta = 0.5$, $\tan \theta = ?$
3. $\tan \theta = 1$, $\cot \theta = ?$
4. $\tan \theta = 8$, $\cot \theta = ?$

11-4 Table of Trigonometric Functions

The ratio of the sides for a specific angle will always be the same for any size triangle. This idea is illustrated in Fig. 11-9. For the tangent ratio of opposite side to adjacent side, the angle of $45°$ shown has $\tan \theta$ equal to $\frac{2}{2}$, $\frac{3}{3}$, or $\frac{4}{4}$. They all equal 1, which is $\tan 45°$. Similarly, the sine or cosine of $45°$ will always be 0.707. Therefore, the angle is specified by its function, or the function defines the angle.

The numerical values are listed in a trigonometric table, such as Table 11-1, for angles from 0 to $90°$. These values can also be obtained with an electronic calculator of the scientific type with trigonometric functions.

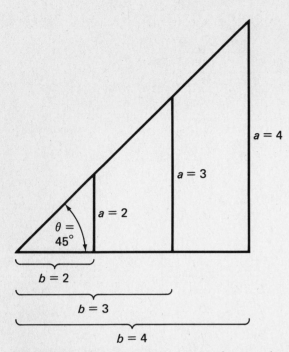

Fig. 11-9 The value of a trigonometric function does *not* depend on the size of the triangle. Here $\theta = 45°$ as tan $45° = \frac{2}{2}, \frac{3}{3}, \frac{4}{4}$, or 1 for any size triangle with equal sides.

Table 11-1 lists angles from 0 to 44° going down the page and then starts at the top again for 45 to 90°. Each angle has its sine, cosine, and tangent listed in a horizontal row. Using the angle of 30° as an example, sin $30° = 0.5000$, cos $30° = 0.8660$, and tan $30° = 0.5774$. As another example, for 60° on the right-hand side of the table, sin $60° = 0.8660$, cos $60° = 0.5000$, and tan $60° = 1.7321$.

Any one angle has the three functions, but we generally use only one at a time. When the problem involves the opposite and adjacent sides without the hypotenuse, then tan θ is useful.

We can determine the functions from the angle or find the angle from a function. For instance, if we know that the angle is 45°, we then can determine that tan $45° = 1$. Or if we know that tan $\theta = 1$, then θ must be 45°.

Note that each function is not in degrees but is just a numerical ratio without any units. It is a pure number, because the units for the sides of the triangle cancel in the ratio. The angle θ itself is in degrees, but its trigonometric functions are not.

Practice Problems 11-K
Answers at End of Chapter

Find the value of the trigonometric functions.

1.	sin 10° =	**9.**	sin 60° =
2.	sin 30° =	**10.**	cos 30° =
3.	cos 60° =	**11.**	tan 18° =
4.	tan 43° =	**12.**	sin 58° =
5.	sin 45° =	**13.**	sin 20° =
6.	cos 45° =	**14.**	cos 70° =
7.	tan 45° =	**15.**	tan 2° =
8.	tan 70° =	**16.**	sin 2° =

Another use of the table is to determine the angle when the value of the function is calculated from the known sides of a triangle. For instance, when the ratio of the opposite side to the hypotenuse is 0.5, then sin θ must be 0.5. Reference to the table will show that sin θ is 0.5 when $\theta = 30°$. Thus, to find an angle when two sides are known, look up the function value in the table and find its corresponding angle.

Example   Find the angles whose function values are as follows:

$$\sin \theta = 0.5; \text{ then } \theta = 30°$$
$$\cos \theta = 0.5; \text{ then } \theta = 60°$$
$$\tan \theta = 1; \text{ then } \theta = 45°$$

Practice Problems 11-L
Answers at End of Chapter

Find angle θ from the function given.

1.	sin $\theta = 0.5446$	**5.**	tan $\theta = 0.0349$
2.	cos $\theta = 0.7193$	**6.**	tan $\theta = 0.8391$
3.	sin $\theta = 0.7071$	**7.**	tan $\theta = 1$
4.	cos $\theta = 0.7071$	**8.**	tan $\theta = 1.1918$

In Table 11-1, sin θ increases from 0 for 0° to 1 for 90°. For 0°, the opposite side is 0 and the sine ratio is 0. For larger angles, the opposite side becomes longer and the sine ratio increases. However, the maximum sine ratio is 1, because no side of the triangle can be longer than the hypotenuse. For 90°, the opposite side can be considered the same as the hypotenuse, for the sine ratio of 1.

Table 11-1 Trigonometric Functions

ANGLE	SIN	COS	TAN	ANGLE	SIN	COS	TAN
0°	0.0000	1.000	0.0000	45°	0.7071	0.7071	1.0000
1	.0175	.9998	.0175	46	.7193	.6947	1.0355
2	.0349	.9994	.0349	47	.7314	.6820	1.0724
3	.0523	.9986	.0524	48	.7431	.6691	1.1106
4	.0698	.9976	.0699	49	.7547	.6561	1.1504
5	.0872	.9962	.0875	50	.7660	.6428	1.1918
6	.1045	.9945	.1051	51	.7771	.6293	1.2349
7	.1219	.9925	.1228	52	.7880	.6157	1.2799
8	.1392	.9903	.1405	53	.7986	.6018	1.3270
9	.1564	.9877	.1584	54	.8090	.5878	1.3764
10	.1736	.9848	.1763	55	.8192	.5736	1.4281
11	.1908	.9816	.1944	56	.8290	.5592	1.4826
12	.2079	.9781	.2126	57	.8387	.5446	1.5399
13	.2250	.9744	.2309	58	.8480	.5299	1.6003
14	.2419	.9703	.2493	59	.8572	.5150	1.6643
15	.2588	.9659	.2679	60	.8660	.5000	1.7321
16	.2756	.9613	.2867	61	.8746	.4848	1.8040
17	.2924	.9563	.3057	62	.8829	.4695	1.8807
18	.3090	.9511	.3249	63	.8910	.4540	1.9626
19	.3256	.9455	.3443	64	.8988	.4384	2.0503
20	.3420	.9397	.3640	65	.9063	.4226	2.1445
21	.3584	.9336	.3839	66	.9135	.4067	2.2460
22	.3746	.9272	.4040	67	.9205	.3907	2.3559
23	.3907	.9205	.4245	68	.9272	.3746	2.4751
24	.4067	.9135	.4452	69	.9336	.3584	2.6051
25	.4226	.9063	.4663	70	.9397	.3420	2.7475
26	.4384	.8988	.4877	71	.9455	.3256	2.9042
27	.4540	.8910	.5095	72	.9511	.3090	3.0777
28	.4695	.8829	.5317	73	.9563	.2924	3.2709
29	.4848	.8746	.5543	74	.9613	.2756	3.4874
30	.5000	.8660	.5774	75	.9659	.2588	3.7321
31	.5150	.8572	.6009	76	.9703	.2419	4.0108
32	.5299	.8480	.6249	77	.9744	.2250	4.3315
33	.5446	.8387	.6494	78	.9781	.2079	4.7046
34	.5592	.8290	.6745	79	.9816	.1908	5.1446
35	.5736	.8192	.7002	80	.9848	.1736	5.6713
36	.5878	.8090	.7265	81	.9877	.1564	6.3138
37	.6018	.7986	.7536	82	.9903	.1392	7.1154
38	.6157	.7880	.7813	83	.9925	.1219	8.1443
39	.6293	.7771	.8098	84	.9945	.1045	9.5144
40	.6428	.7660	.8391	85	.9962	.0872	11.43
41	.6561	.7547	.8693	86	.9976	.0698	14.30
42	.6691	.7431	.9004	87	.9986	.0523	19.08
43	.6820	.7314	.9325	88	.9994	.0349	28.64
44	.6947	.7193	.9657	89	.9998	.0175	57.29
				90	1.0000	.0000	∞

Practice Problems 11-M
Answers at End of Chapter

Find sin θ for the following angles.

1.	0°	**5.**	60°
2.	20°	**6.**	70°
3.	30°	**7.**	80°
4.	45°	**8.**	90°

The values for cos θ start from 1 as a maximum value for 0°. The reason is that for 0°, the adjacent side can be considered the same as the hypotenuse and the cosine ratio is unity. Then cos θ decreases for larger angles as the adjacent side becomes smaller. At 90°, the adjacent side can be considered zero and cos θ = 0.

Practice Problems 11-N
Answers at End of Chapter

Find cos θ for the following angles.

1.	0°	**5.**	60°
2.	20°	**6.**	70°
3.	30°	**7.**	80°
4.	45°	**8.**	90°

The tangent function is probably used most often, to find the phase angle in ac circuits. The reason is that tan θ involves only the perpendicular sides, which correspond to resistance and reactance.

The values of tan θ increase without limit as angle θ increases toward 90°, because the opposite side becomes longer. However, tan θ values should be considered in two parts, below and above 45°:

1. Up to 45°, tan θ increases from 0 to 1. At 45°, the opposite and adjacent sides are equal, for a ratio of 1 for tan θ. A subdivision here is tangent values of less than 0.1 for angles less than 5.7°.
2. Above 45°, tan θ is more than 1 because the opposite side is longer than the adjacent side. A subdivision here is tangent values of more than 10 for angles above 84.3°. At 90°, the tangent ratio is infinite (∞), as the opposite side then is infinitely long.

The main thing to remember about the tangent function is that when the opposite side is longer, tan θ is more than 1; when the adjacent side is longer, tan θ is less than 1.

Practice Problems 11-O
Answers at End of Chapter

Find tan θ for the following angles.

1.	0°	**6.**	72°
2.	6°	**7.**	84°
3.	30°	**8.**	88°
4.	45°	**9.**	5.7°
5.	60°	**10.**	84.3°

Practice Problems 11-P
Answers at End of Chapter

Calculate tan θ and find θ from the sides in the following examples. (a = opposite side, b = adjacent side.)

1.	$b = 4, a = 1$	**6.**	$b = 4, a = 6$
2.	$b = 4, a = 2$	**7.**	$b = 4, a = 8$
3.	$b = 4, a = 3$	**8.**	$b = 4, a = 40$
4.	$b = 4, a = 4$	**9.**	$b = 10, a = 1$
5.	$b = 4, a = 5$	**10.**	$b = 1, a = 10$

11-5 Angles of More Than 90°

The full circle of 360° is divided into four quadrants of 90° each, as shown in Fig. 11-10. In the counterclockwise direction, the quadrants are as follows:

I	0 to 90°
II	90 to 180°
III	180 to 270°
IV	270 to 360° (or 0°)

After 360° or any multiple of 360°, the angles just repeat the values from 0°.

To use the trigonometric functions for angles in quadrants II, III, and IV, these values can be converted to equivalent angles in quadrant I. The following rules apply for θ greater than 90°:

In quadrant II, use 180° − θ
In quadrant III, use θ − 180°
In quadrant IV, use 360° − θ

Note that the conversions are with respect to the horizontal axis only, using either 180° or 360° as the refer-

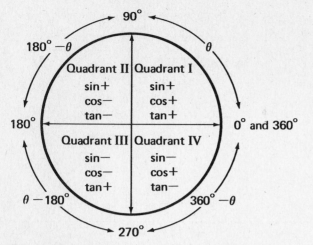

Fig. 11-10 Angles in the four quadrants of a circle.

ence. This way the obtuse angle is always subtracted from a larger angle.

Example	Convert 135° to an angle of less than 90°.
Answer	The angle is in quadrant II. Therefore, using the formula,

$$\text{Angle} = 180 - \theta$$
$$= 180 - 135$$
$$= 45°$$

Practice Problems 11-Q
Answers at End of Chapter

Convert to an acute angle of less than 90°.

1.	160°	**5.**	210°
2.	200°	**6.**	330°
3.	240°	**7.**	315°
4.	150°	**8.**	390°

It is also necessary to consider the sign or polarity of the function in each quadrant.

All the functions are positive in quadrant I.

The sine is also positive in quadrant II, where the vertical ordinate is still positive. However, the sine is negative in quadrants III and IV, where the vertical axis (ordinate) is negative.

The cosine is negative in quadrants II and III, where the horizontal axis (abscissa) is negative, but positive in quadrants IV and I.

The tangent alternates in sign through the quadrants, positive in quadrants I and III. In quadrant III, both sides are negative, which makes the tangent positive. Examples of conversion for tan θ are as follows:

In quadrant II,

$$\tan 120° = -\tan (180° - 120°)$$
$$= -\tan 60° = -1.7321$$

In quadrant III,

$$\tan 240° = \tan (240° - 180°)$$
$$= \tan 60° = +1.7321$$

In quadrant IV,

$$\tan 300° = -\tan (360° - 300°)$$
$$= -\tan 60° = -1.7321$$

Example	Find the value of the sine, cosine, and tangent for an angle of 315°, in quadrant IV.
Answer	$\sin 315° = -\sin (360° - 315°)$

$$\quad = -\sin 45° = -0.7071$$
$$\cos 315° = +\cos 45° = +0.7071$$
$$\tan 315° = -\tan 45° = -1.000$$

Figure 11-11 shows examples of angles in each of the four quadrants.

Practice Problems 11-R
Answers at End of Chapter

Find sin θ, cos θ, and tan θ with the correct sign for the following obtuse angles.

1.	$\theta = 160°$	**3.**	$\theta = 210°$
2.	$\theta = 200°$	**4.**	$\theta = 330°$

11-6 Radian Measure of Angles

In circular measure it is convenient to use a specific unit angle called the *radian* (abbreviated rad). Its con-

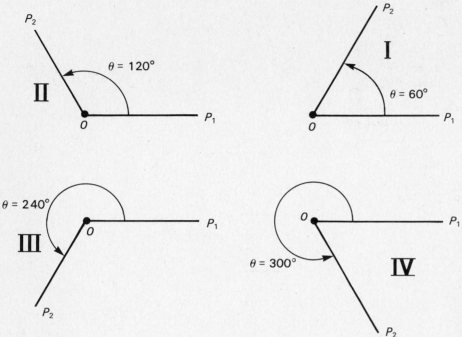

Fig. 11-11 Examples of an angle in each of the four quadrants. The 60° in quadrant I corresponds to 120°, 240°, and 300°.

venience is due to the fact that a radian is the angular part of a circle that includes an arc equal to the radius, as shown in Fig. 11-12 on page 104.

The circumference of a circle is $2\pi r$ in length, where r is the radius. The Greek letter π is the ratio of the circumference to the diameter for any circle. Then π is a constant value at 3.14, and $2\pi = 6.28$

Since the circumference is $2\pi r$ and since one radian uses one length of r, the complete circle includes 2π rad. A circle also is 360°. Therefore, 2π rad = 360°. Or one radian is 360° divided by 2π. This division gives

$$1 \text{ rad} = \frac{360°}{6.28} = 57.3°$$

The quadrants of a circle can be considered conveniently in radian angles instead of degrees. The corresponding values are shown in the table below:

Quadrant			I			II	III	IV
Degrees	0	30	45	60	90	180	270	360
Radians	0	$\dfrac{\pi}{6}$	$\dfrac{\pi}{4}$	$\dfrac{\pi}{3}$	$\dfrac{\pi}{2}$	π	$\dfrac{3\pi}{2}$	2π

In general, to convert degrees into radians, divide the angles in degrees by 57.3°/rad.

Example	Express 86° in radians.
Answer	Angle in rad = 86° ÷ 57.3°/rad $= \dfrac{86° \text{ rad}}{57.3°}$ $= 1.5 \text{ rad}$

Note that the degrees cancel.

To convert radians to degrees, multiply by 57.3°/rad.

Example	Convert 1.5 rad to degrees.
Answer	Angle in degrees = 1.5 rad × 57.3°/rad $= 1.5 × 57.3°$ $= 86°$

Note that the radians cancel.

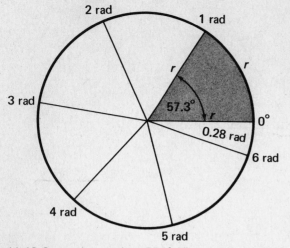

Fig. 11-12 One radian angle is 57.3°. The complete circle includes 2π rad or 6.28 rad.

Practice Problems 11-S
Answers at End of Chapter

Convert to radians. Write the answer in terms of π.

1. $45° =$ **4.** $270° =$
2. $90° =$ **5.** $360° =$
3. $180° =$ **6.** $720° =$

Practice Problems 11-T
Answers at End of Chapter

Convert to degrees.

1. $\dfrac{\pi}{2}$ rad $=$ **5.** $\dfrac{3\pi}{2}$ rad $=$
2. 1.75 rad $=$ **6.** 5.24 rad $=$
3. π rad $=$ **7.** 5.76 rad $=$
4. 3.5 rad $=$ **8.** 2π rad $=$

11-7 Special Values of Angles

The trigonometric functions of a few angles are listed in Table 11-2 because they illustrate the general idea of how the values vary. The few functions here can be considered a miniature version of the complete trigonometric table in Table 11-1. Specifically, the angles for 30°, 45°, and 60°, with their functions, are shown in Fig. 11-13.

Consider the angle of 45°. Tan 45° is 1, as the opposite and adjacent sides are equal. For angles larger than

45°, tan θ is greater than 1. Tan θ is less than 1 for angles less than 45°.

Furthermore, at 45° cos θ and sin θ are equal at 0.707. Sin θ increases toward 1 for angles up to 90°, while cos θ decreases toward zero for larger angles.

For 30°, sin θ is exactly one-half or 0.5. For 60°, sin θ increases to 0.866.

Cos θ for 30° is 0.866. For 60°, cos θ decreases to exactly 0.5.

The angles of 30° and 60° are complementary, as 30° + 60° = 90°. Note that sin 30° at 0.5 is the same as cos 60° at 0.5. For any two complementary angles, sin θ for one angle is equal to cos θ for the other angle.

Practice Problems 11-U
Answers at End of Chapter

1.	sin 0° =	**6.**	cos 0° =
2.	sin 30° =	**7.**	cos 30° =
3.	sin 45° =	**8.**	cos 45° =
4.	sin 60° =	**9.**	cos 60° =
5.	sin 90° =	**10.**	cos 90° =

The way the sine function varies is illustrated graphically in Fig. 11-14a. Values at 30°, 45°, 60°, and 90° are marked on the curve for quadrant I. Remember that angles of more than 90° can be converted to equivalent acute angles for finding the trigonometric functions, using the methods explained in Sec. 11-5.

The graphs in Fig. 11-14 are known as *sinusoidal waveforms*. The sine wave of Fig. 11-14a corresponds to the values of sin θ from 0° to 360° for the complete circle. It is a periodic waveform, as the values are repeated periodically every 360°. The value at 390°, for instance, is the same as 390 − 360 = 30°.

An important characteristic of the sine wave is that the values have a faster rate of change near the zero

Table 11-2 Trigonometric Functions for Special Angles

Angle θ	sin θ	cos θ	tan θ
0°	0	1	0
30°	0.5	0.866	0.577
45°	0.707	0.707	1.0
60°	0.866	0.5	1.732
90°	1.0	0	∞

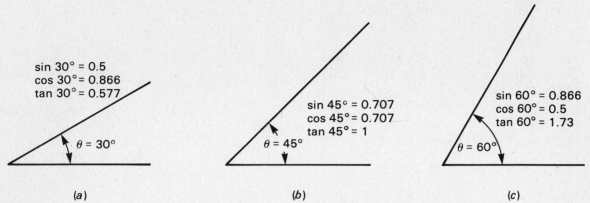

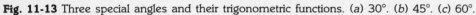

Fig. 11-13 Three special angles and their trigonometric functions. (a) 30°. (b) 45°. (c) 60°.

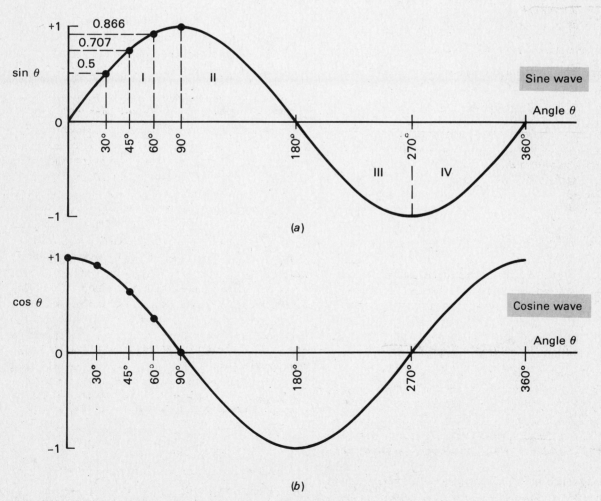

Fig. 11-14 Sinusoidal waveforms. (a) Sine wave. (b) Cosine wave.

axis, compared with the maximum and minimum values at +1 and -1. The curve rises to 0.5, or one-half the maximum amplitude, at 30°, which is only one-third of 90°. In the 30° between 60° and 90°, near the top, the curve rises only by the difference between 1.0 and 0.866, which equals 0.134.

The values in quadrant II of the sine wave are an exact mirror image of those in quadrant I. In other words, the curve starts down gradually, near the top, then has a faster decrease in values near the zero axis. Furthermore, the negative part of the curve, for quadrants III and IV, is exactly the same as the positive part, for quadrants I and II, but with increasing negative values.

The cosine wave in Fig. 11-14b has the same values as the sine wave in Fig. 11-14a but displaced by 90°. In other words, a cosine wave and a sine wave are in quadrature, meaning that they differ by 90°. As an example, for 0°, sin θ is 0 and cos θ is 1. Sin θ is equal to 1 at an angle 90° away.

Practice Problems 11-V
Answers at End of Chapter

Refer to the curves in Fig. 11-14.

1.	sin 0° =	**5.**	sin 270° =
2.	cos 0° =	**6.**	cos 270° =
3.	sin 60° =	**7.**	cos 60° =
4.	sin 120° =	**8.**	cos 120° =

Another important angle is 5.7°. This is the dividing line for small angles that can be considered practically zero. For angles 5.7° and smaller, sin θ and tan θ are practically equal. The reason is that the ratio a/b for tan θ and a/c for sin θ are almost the same, as the hypotenuse is very little longer than side b (Fig. 11-15). Specifically, at 5.7°, the value for sin θ or tan θ is 0.1. For smaller angles down to 0.57°, both sin θ and tan θ have values with one zero after the decimal point. For in-

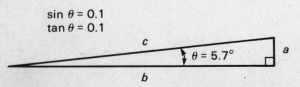

Fig. 11-15 Angle θ of 5.7°. This value is the dividing line for small angles where sin θ and tan θ are practically the same, both equal to 0.1.

Table 11-3 Increasing Values for sin θ and tan θ

Angle	sin θ	tan θ
0°	0	0
0.57°	0.01	0.01
5.7°	0.1	0.1
30°	0.5	0.577
45°	0.707	1
60°	0.866	1.732
84.3°	0.995	10.0
90°	1.0	∞

stance, at 4°, sin θ and tan θ are equal at approximately 0.07.

The way that sin θ and tan θ increase from 0° to 90° is shown in Table 11-3. Notice that tan θ increases from 0.01 through 0.1 to 1 at 45°. Also, tan θ is 10 for 84.3°, which is 5.7° less than 90°. Angles more than 84.3° have tan θ with two places before the decimal point. For example, tan 88° is 28.6.

Practice Problems 11-W
Answers at End of Chapter

1.	sin 3° =	**5.**	sin 86° =
2.	tan 3° =	**6.**	tan 86° =
3.	sin 30° =	**7.**	tan 45° =
4.	tan 30° =	**8.**	sin 45° =

11-8 The Inverse Trigonometric Functions

When we say sin 30° is 0.5, it follows that the angle with a sine value of 0.5 must be 30°. Thus 0.5 is an inverse trigonometric function of 30°. Inverse functions are commonly used for the sine, cosine, and tangent. The method of indicating these inverse functions is as follows:

Function	Inverse Function
sin θ	$\sin^{-1} \theta$ or arcsin θ
cos θ	$\cos^{-1} \theta$ or arccos θ
tan θ	$\tan^{-1} \theta$ or arctan θ

The exponent of -1 or the word *arc* indicates an inverse function. As an example, arcsin 0.5 or

$\sin^{-1} 0.5$ means 30°. Similarly, $\cos^{-1} 0.5$ indicates 60°; also, arctan 1 means an angle of 45°.

Practice Problems 11-X
Answers at End of Chapter

Give the angles for the following inverse trigonometric functions.

1.	$\tan^{-1} 0.1$	**6.**	arctan 10
2.	$\sin^{-1} 0.1$	**7.**	arcsin 0.707
3.	$\cos^{-1} 0.707$	**8.**	arcsin 1
4.	$\tan^{-1} 1$	**9.**	arctan 1.73
5.	$\sin^{-1} 0.5$	**10.**	arcsin 0.1

11-9 Using the Electronic Calculator for Trigonometric Functions

The scientific calculator usually has provisions for calculating $\sin \theta$, $\cos \theta$, $\tan \theta$, and their inverse functions. This method is much easier than using a trigonometric table. First, make sure the calculator is set for the angular mode you want—degrees, radians, or grads. Usually, degrees are used with decitrig notation. In other words, specify the angle in degrees and decimal fractions of one degree, such as 24.6.

The angular mode is chosen with the ⌈DRG⌉ key. The D is for degrees, the R is for radians, and the G is for grads. The mode changes from D to R to G as you press the key. Which mode is being used will be indicated on the visual display.

To find the value of the trigonometric function, just punch in the angle as a decimal number. Then push ⌈SIN⌉, ⌈COS⌉, or ⌈TAN⌉ and the value of the function will be displayed. For 24.6°, as an example, punch in 24.6 as a decimal number. Push ⌈SIN⌉ and the displayed value is 0.416. Also, ⌈COS⌉ can be used for $\cos \theta$ or ⌈TAN⌉ for $\tan \theta$. The results for this example of 24.6° are

$$\sin 24.6° = 0.416$$
$$\cos 24.6° = 0.909$$
$$\tan 24.6° = 0.458$$

Furthermore, the calculator can give the inverse trigonometric functions. Usually, a separate key must be pressed before the function key, either an "invert" key or a "second function" key. As an example, find $\sin^{-1}$ 0.416. Punch in the value as a decimal number, push ⌈INV⌉ or ⌈2ndF⌉ next, then push ⌈SIN⌉ and the inverse function will be displayed. The results for this example are

$$\sin^{-1} 0.416 = 24.6°$$
$$\cos^{-1} 0.909 = 24.6°$$
$$\tan^{-1} 0.458 = 24.6°$$

The inverse function is often used for tangent values because in many cases the opposite and adjacent sides are known but the hypotenuse is not. For instance, suppose that the opposite side is 12 and the adjacent side is 10. Then $\tan \theta$ is ¹²⁄₁₀, or arctan 1.2 is equal to θ. With the calculator, for arctan 1.2 or $\tan^{-1}$ 1.2, θ is equal to 50.2°. Note that this angle must be more than 45° because $\tan \theta$ is more than 1.

An application of this example can be to find the phase angle in a series ac circuit. Let X_L be 12 Ω with R of 10 Ω, as in Fig. 11-6a. Then

$$\tan \theta = \frac{12}{10} = 1.2$$
$$\text{arctan } 1.2 = 50.2°$$

The phase angle of 50.2° is the angular difference in time between the sine wave of applied voltage and the sine wave of current in the circuit.

Review Problems
Answers to Odd-Numbered Problems at Back of Book

Use an electronic calculator to find the following values.

1.	cos 0°	**11.**	sin 19.4°
2.	sin 30°	**12.**	tan 42°
3.	sin 45°	**13.**	tan 48°
4.	tan 45°	**14.**	$\tan^{-1}$ 0.1
5.	cos 60°	**15.**	$\tan^{-1}$ 0.9
6.	cos 30°	**16.**	$\tan^{-1}$ 1.0
7.	sin 5.7°	**17.**	$\tan^{-1}$ 1.111
8.	tan 5.7°	**18.**	$\tan^{-1}$ 0.75
9.	tan 84.3°	**19.**	$\tan^{-1}$ 1.33
10.	sin 24.6°	**20.**	$\tan^{-1}$ 2.0

Answers to Practice Problems

11-A
1. 90°
2. 180°
3. 270°
4. 36°
5. 120°
6. 60°
7. 72°
8. 360°

11-B
1. Acute
2. Acute
3. Right
4. Obtuse
5. Acute
6. Obtuse
7. Acute
8. Acute

11-C
1. 60°
2. 30°
3. 45°
4. 73°
5. 70°
6. 37°
7. 50°
8. 40°

11-D
1. 40°
2. 20.5°
3. 70°
4. 17°
5. 120°
6. 50°
7. 5°
8. −6°
9. 60°
10. 30°

11-E
1. 10
2. 10
3. 7.07
4. 4.47
5. 4.47
6. 10.05
7. 10.05
8. 14.14
9. 6.4
10. 8.25

11-F
1. $b = 10$
2. $a = 10$
3. $a = 3$

4. $b = 4$
5. $b = 4$
6. $a = 4$
7. $a = 4$
8. $b = 8$

11-G
1. $Z = 10 \ \Omega$
2. $Z = 10 \ \Omega$
3. $Z = 7.07 \ \Omega$
4. $Z = 7.07 \ \Omega$
5. $Z = 4.25 \ \Omega$
6. $Z = 6.4 \ \Omega$
7. $Z = 10.05 \ \Omega$
8. $Z = 10.05 \ \Omega$
9. $Z = 5 \ \Omega$
10. $Z = 5 \ \Omega$

11-H
1. $\phi = 60°$
2. $\phi = 30°$
3. $\theta = 45°$
4. $\theta = 70°$
5. $\phi = 85°$
6. $\theta = 85°$
7. $\phi = 37°$
8. $\theta = 77.5°$

11-I
1. 1
2. 0.5
3. 0.8
4. 0.8
5. 0.6
6. 0.75
7. 1
8. 1
9. 1
10. 0.707

11-J
1. 0.5
2. 2
3. 1
4. 0.125

11-K
1. 0.1736
2. 0.5
3. 0.5
4. 0.9325
5. 0.7071
6. 0.7071
7. 1
8. 2.7475
9. 0.8660
10. 0.8660

11. 0.3249
12. 0.8480
13. 0.342
14. 0.342
15. 0.035
16. 0.035

11-L
1. 33°
2. 44°
3. 45°
4. 45°
5. 2°
6. 40°
7. 45°
8. 50°

11-M
1. 0
2. 0.3420
3. 0.5
4. 0.7071
5. 0.8660
6. 0.9397
7. 0.9848
8. 1

11-N
1. 1
2. 0.9397
3. 0.8660
4. 0.707
5. 0.5
6. 0.3420
7. 0.1736
8. 0

11-O
1. 0
2. 0.1051
3. 0.5774
4. 1
5. 1.7321
6. 3.0777
7. 9.5144
8. 28.64
9. 0.1
10. 10.0

11-P
1. 14°
2. 26.6°
3. 36.9°
4. 45°
5. 51.3°
6. 56.3°
7. 63.4°

8. 84.3°
9. 84.3°
10. 5.7°

11-Q 1. 20°
2. 20°
3. 60°
4. 30°
5. 30°
6. 30°
7. 45°
8. 30°

11-R 1. $\sin \theta = +0.3420$
$\cos \theta = -0.9397$
$\tan \theta = -0.3640$
2. $\sin \theta = -0.3420$
$\cos \theta = -0.9397$
$\tan \theta = +0.3640$
3. $\sin \theta = -0.5000$
$\cos \theta = -0.8660$
$\tan \theta = +0.5774$
4. $\sin \theta = -0.5000$
$\cos \theta = +0.8660$
$\tan \theta = -0.5774$

11-S 1. $\frac{\pi}{4}$ rad

2. $\frac{\pi}{2}$ rad
3. π rad
4. $\frac{3\pi}{2}$ rad
5. 2π rad
6. 4π rad

11-T 1. 90°
2. 100°
3. 180°
4. 200°
5. 270°
6. 300°
7. 330°
8. 360°

11-U 1. 0
2. 0.5
3. 0.707
4. 0.866
5. 1.0
6. 1.0
7. 0.866
8. 0.707
9. 0.5
10. 0

11-V 1. 0
2. 1
3. 0.866
4. 0.866
5. −1
6. 0
7. 0.5
8. −0.5

11-W 1. 0.052
2. 0.052
3. 0.5
4. 0.577
5. 0.998
6. 14.30
7. 1.0
8. 0.707

11-X 1. 5.7°
2. 5.7°
3. 45°
4. 45°
5. 30°
6. 84.3°
7. 45°
8. 90°
9. 60°
10. 5.7°

12 COMPUTER MATHEMATICS

The first arithmetic we were taught consisted of ten symbols called digits: 0, 1, 2, 3, 4, 5, 6, 7, 8, and 9. These digits make up the decimal number system, the prefix deci meaning ten. The first eleven chapters of this book deal with operations in the decimal number system. Other number systems are also possible. With the advent of computers, number systems based on two, eight, and sixteen digits became very useful.

The binary number system consists of only two digits: 0 and 1.

The octal number system consists of eight digits: 0, 1, 2, 3, 4, 5, 6, and 7.

The hexadecimal number system consists of sixteen digits. However, since we have only ten numerical symbols, the hexadecimal system uses letters also. The sixteen digits in the hexadecimal system are 0, 1, 2, 3, 4, 5, 6, 7, 8, 9, A, B, C, D, E, and F.

Binary numbers are widely used in applications that deal with a choice of one of two conditions, for example, on-off or high-low or yes-no or true-false. Boolean algebra (named after George Boole, a British logician) and truth tables are used to solve problems that deal with choice and logic such as found in the field of computers and controls. The main features of computer mathematics are explained in the following sections:

12-1 Binary Numbers

Binary numbers use only the two digits 0 and 1 for different place counts. For instance, the binary number $(1001)_2$ corresponds to the decimal number $(9)_{10}$. The subscripts 2 and 10 indicate the respective bases of the numbers. The system of binary numbers is used in digital electronics because the circuits are either on or off, or high or low. The binary value of 1 can be called the HIGH state and the binary value of 0 the LOW state. In this way, binary notation, such as 1001, can indicate the HIGH and LOW levels in a group of voltage pulses. The train of pulses is a digital signal, containing information about the sequence of HIGH and LOW levels.

The binary numbers are usually written in groups of four or eight. Each 0 or 1 is called a *bit*. A group of four bits is a *nibble*. Eight bits make a *byte*.

Practice Problems 12-A
Answers at End of Chapter

1. The subscript 2 indicates $(1001)_2$ is a number in the ____ number system.
2. If 1 represents ON in a binary system, 0 represents ____ .
3. The train of pulses 1001 represents a ____ signal.

12-2 Counting with Binary Numbers

Consider the binary number $(1101)_2$. Each 1 or 0 is a place count. Successive place values increase in multiples of 2, from right to left, as shown in Fig. 12-1. Therefore, the powers of 2 must be memorized for binary numbers. These values are

$$2^0 = 1 \quad\quad 2^2 = 4 \quad\quad 2^4 = 16 \quad\quad 2^6 = 64$$
$$2^1 = 2 \quad\quad 2^3 = 8 \quad\quad 2^5 = 32 \quad\quad 2^7 = 128$$

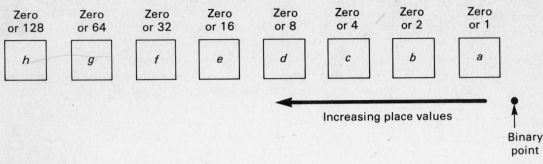

Fig. 12-1 Place values for binary numbers.

As an example, 2^3 is $2 \times 2 \times 2$, using 2 as a multiplying factor three times, for the value of 8. As a result, the binary number $(1101)_2$ is equal to $8 + 4 + 0 + 1$ for the decimal value of $(13)_{10}$.

Since binary numbers have only two values, either 1 or 0, for each place count, the base or *radix* of the system is 2. This corresponds to the base, or radix, of 10 for decimal numbers. The highest digit that can be used in each place is one less than the radix. Then the highest digit for any place value in the binary system is 1, just as the highest digit for any place in the decimal system is 9. There is no such thing as 10 in one decimal place because it has a carry to the next place. Similarly, you cannot have a 2 in binary numbers, as the value is binary 0 with a carry of 1 to the next higher place.

In Fig. 12-1, the place at the right to the left of the binary point is the least significant figure. The place at the left with the highest value is the most significant. For all the places, the count can only be 0 or 1, which means zero or the maximum count for that place. The maximum count for each place is as follows:

Place a is zero or 1. This is the 1s place.
Place b is zero or $2^1 = 2$. This is the 2s place.
Place c is zero or $2^2 = 4$. This is the 4s place.
Place d is zero or $2^3 = 8$. This is the 8s place.
Place e is zero or $2^4 = 16$. This is the 16s place.
Place f is zero or $2^5 = 32$. This is the 32s place.
Place g is zero or $2^6 = 64$. This is the 64s place.
Place h is zero or $2^7 = 128$. This is the 128s place.

Only eight places are illustrated, but more places can be used for bigger numbers. Note that the place values increase in multiples of 2 in the order 2, 4, 8, 16, 32, 64, 128. The next higher place value would be $2 \times 128 = 256$. Actually, the binary place values are rela-

tively easy because the place either has its maximum count or is zero.

The way these place values apply to binary numbers for 4-digit and 8-digit numbers is shown by these examples:

$$(0001)_2 = 1 \qquad (0001\ 0000)_2 = 16$$
$$(0010)_2 = 2 \qquad (0010\ 0000)_2 = 32$$
$$(0100)_2 = 4 \qquad (0100\ 0000)_2 = 64$$
$$(1000)_2 = 8 \qquad (1000\ 0000)_2 = 128$$

Note that the binary point, omitted here, is assumed to be after the last digit on the right, just as in decimal numbers. The base for these binary numbers is indicated by the subscript 2 for the group of digits, but this subscript may be omitted when the number is obviously in binary form.

These examples all have a count, indicated by binary 1, in only one of the places. For numbers with a count in more than one place, just add the count for each place. As examples,

$$(1111)_2 = 8 + 4 + 2 + 1 = (15)_{10}$$
$$(1010)_2 = 8 + 0 + 2 + 0 = (10)_{10}$$

The process of adding the binary count for each place is converting the binary number to the equivalent decimal number. The decimal value is indicated by the subscript 10, if it is necessary to emphasize that it is a decimal number.

It should be noted that places to the right of the binary point can also be used, if necessary, for values less than decimal 1. Then each place decreases in multiples of ½, going to the right from the binary point. As examples, $(0.1)_2$ is $(\frac{1}{2})_{10}$ and $(0.01)_2$ is $(\frac{1}{4})_{10}$, for fractional values.

Practice Problems 12-B
Answers at End of Chapter

Convert the following binary values to decimal numbers.

1.	0000	6.	1100
2.	0001	7.	1110
3.	0010	8.	1111
4.	0100	9.	0111
5.	1000	10.	0011

12-3 Binary Addition

In binary arithmetic, addition is the main process. While binary subtraction can also be performed, the process involves addition. This method is explained in Sec. 12-4. Multiplication and division of binary numbers can be done by a special sequence of operations called an *algorithm*, which uses addition and complements of numbers.

For addition of binary 1 and 0 for any place count, all the possibilities are

$$0 + 0 = 0$$
$$1 + 0 = 1$$
$$0 + 1 = 1$$
$$1 + 1 = 0 \text{ and carry } 1$$

The first three additions are the same as in decimal arithmetic. Note that $1 + 0$ is the same as $0 + 1$. For the fourth case of binary addition, though, $1 + 1$ is equal to 0 plus a carry of 1. We cannot have the digit 2 in binary numbers. The decimal number 2 corresponds to binary 0 plus a carry of 1 to the next higher place, which is written as 10.

Example   Add 1001 and 0110.

Answer   Line up the binary point, just as in addition of decimal numbers. Then

$$
\begin{array}{ll}
(1001)_2 & (9)_{10} \\
+ (0110)_2 & + (6)_{10} \\
\hline
(1111)_2 & (15)_{10}
\end{array}
$$

The binary sum 1111 is the decimal value of $8 + 4 + 2 + 1 = 15$.

This binary example does not have a carry of 1 in any of the columns to be added. However, an overflow will result when there is an addition of $1 + 1$. The result is a *carry-out* of 1 to the next higher place column, to the left. This column then has a *carry-in* of 1 to be added. The possibilities for the carry of 1 are

$$1 + 1 = 0 \text{ plus carry-out of } 1$$
$$1 + 1 + \text{carry-in of } 1 = 1 \text{ plus carry-out of } 1$$
$$1 + 0 + \text{carry-in of } 1 = 0 \text{ plus carry-out of } 1$$
$$0 + 0 + \text{carry-in of } 1 = 1 \text{ without any carry-out}$$

A carry-out of 1 is the overflow from the column being added. A carry-in is 1 added to the column to the left with the next higher place value. The possibility of a carry is the reason why the columns are added from right to left, from place a at the binary point through places $b, c, d, e, f, g,$ and h as shown in Fig. 12-1. The carry-in must go to the next place that has a higher count, not a lower count.

When the carry-in of 1 is to the 2s column, the carry is the equivalent of adding decimal 2. In the 4s column, the carry-in of binary 1 adds decimal 4. Also, a carry-in of binary 1 in the 8s column adds decimal 8, and so on for all the columns. A problem in binary addition with examples of the carry follows.

Example   Add $(111)_2$ and $(101)_2$.

Answer
$$
\begin{array}{ll}
\quad cba & \\
(111)_2 & (7)_{10} \\
+ (101)_2 & + (5)_{10} \\
\hline
(1100)_2 & (12)_{10}
\end{array}
$$

For the first column a at the right, $1 + 1 = 0$ carry 1 to column b.

In column b, $1 + 0 +$ carry-in of 1 is the same as $1 + 1$, which equals 0 with carry-out of 1 to column c.

This carry-in of 1 makes column c $1 + 1 + 1$, which equals 1 with carry-out of 1 to the next and last column.

An extra place is needed for the sum because the last column has a carry-out. This answer of binary 1100 is equal to the decimal addition of $8 + 4 + 0 + 0 = 12$.

Answers at End of Chapter

Do the following binary additions. Give the sum in binary and decimal form.

1.	1001 + 0110 =	**6.**	1111 + 0111 =
2.	0001 + 1000 =	**7.**	1111 + 0001 =
3.	0010 + 0001 =	**8.**	0111 + 0111 =
4.	0101 + 1010 =	**9.**	1100 + 1101 =
5.	1001 + 1001 =	**10.**	1111 + 1111 =

12-4 Subtraction by Adding the Complement

The method of complementation is used to find the difference between two binary numbers. The complement of a number is related to the radix (R) or base of the number system. For instance, in the decimal system, with R of 10, the R complement of 6 is 4, as 6 + 4 = 10. Another technique uses the R − 1 complement. In the decimal system, R − 1 is equal to 9. Therefore the R − 1 complement of 6 is 3, as 6 + 3 = 9. In the binary system, the R − 1 complement is called the 1s complement, as R is 2 and 2 − 1 is equal to 1.

The 1s complement for binary numbers is very neat because 1 has the complement of 0 and 0 has the complement of 1. For instance, the 1s complement of 1110 is 0001, where each digit is replaced by its 1s complement.

Answers at End of Chapter

Give the 1s complement of the following binary numbers.

1.	0000	**5.**	1001
2.	1111	**6.**	0110
3.	0101	**7.**	0011
4.	1010	**8.**	1100

Subtraction with binary numbers can be done by adding the complement of the subtrahend, which is the number to be subtracted. This process corresponds to subtraction in decimal arithmetic by changing the sign of the subtrahend and adding. For binary numbers using the 1s complement for subtraction, the rules are:

1. Convert the subtrahend to its complement by changing 0 to 1 and 1 to 0 for all the digits.
2. Add the complement to the original minuend, which is the number from which the subtrahend is to be subtracted.
3. If there is a carry-out of 1 from the last column, for the most significant place at the left, move this carry around to the least significant place at the right and add the 1 in this column.
4. If there is no carry-out for the last column, the answer for the subtraction is negative, meaning that the subtrahend is larger than the minuend. In this case, take the complement of all the digits in the sum and give this answer a negative sign.

A few examples will illustrate this method. Let the problem be to subtract 0011 (the subtrahend) from 1001 (the minuend). In decimal form, the problem is to subtract 3 from 9. So, the answer should be 6. With binary numbers, the 1s complement of the subtrahend 0011 is 1100. Add this complement to 1001.

```
  1001 (original minuend)
+ 1100 (complement)
  ┌ 0101
  └──→ 1 (carry around)
  0110 (answer)
```

The answer, 0110, is equal to 6. Note how the carry-out of 1 is moved from the column at the left to the column at the right.

As another example, subtract 1001 from 0011. In decimal form this problem is 3 − 9, for the answer of −6. The numbers are the same as in the previous problem, but here the larger number is subtracted from the smaller number. In binary form, the subtrahend of 1001 has the 1s complement of 0110. Add this complement to 0011 to get the complement of the sum with minus sign.

```
   0011
 + 0110 (complement)
   1001 (sum)
  −0110 (answer)
```

Adding the complement gives 1001, but since there is no carry-out from the first column at the left, the answer is negative. Therefore, it is necessary to find the complement of the sum, which is 0110. Add a negative sign to this number for the answer −0110 or −6.

Practice Problems 12-E
Answers at End of Chapter

Do the following subtractions and give the answer in binary and decimal form.

1.	$1001 - 0110 =$	**5.**	$0001 - 0011 =$
2.	$0110 - 0011 =$	**6.**	$1111 - 0001 =$
3.	$1111 - 0111 =$	**7.**	$1011 - 0111 =$
4.	$0110 - 0110 =$	**8.**	$0111 - 1011 =$

12-5 Converting Decimal Numbers to Binary Numbers

A common method used to convert a decimal number to its equivalent binary number is to divide the decimal number by 2 in successive steps until the quotient is zero. This method is called *double-dabble* because of the division by 2. It is also called the remainder method. The remainders are the digits in each place of the equivalent binary number. The first division and its remainder provide the binary digit for the least significant place at the right. The following remainders fill in each of the binary places, from right to left. As an example, we can convert decimal 8 to binary form. Then

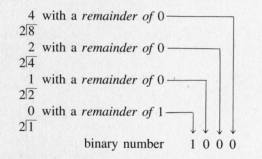

Only the remainders of 1 or 0 are used for the resulting binary number of 1000. This is $8 + 0 + 0 + 0 = 8$.

The first step, 8 divided by 2, has a remainder of 0 for the place next to the binary point. Next, 4 is divided by 2, which also has a remainder of 0 for the next place to the left. Also, 2 divided by 2 has a quotient of 1, but the remainder is 0 for the third binary place. In the last division, 1 is divided by 2. The quotient is actually ½ or 0.5, but this is stated as 0 + remainder of 1 in order to have a separate remainder. As a result, the fourth place in this binary number is 1 for the complete value of $(1000)_2$.

It is important to continue the divisions until the quotient is zero. Otherwise, a 1 might be missing from the

last place to the left in the binary number, which is the most significant place.

When we consider the binary places in the same order as the successive divisions, the remainders of 1000 for the binary number are produced from right to left. The progression of 1 or 0 digits is from the least significant to the most significant place. For the opposite direction, when we consider the binary places starting with the last division, the remainders of 1 or 0 for the binary number go from left to right. Then the progression of 1 or 0 digits is from the most significant to the least significant place.

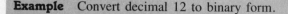

> **Example** Convert decimal 12 to binary form.
>
> **Answer**
>
> $$12 \div 2 = 6 \text{ with a } remainder \text{ } of \text{ } 0$$
> $$6 \div 2 = 3 \text{ with a } remainder \text{ } of \text{ } 0$$
> $$3 \div 2 = 1 \text{ with a } remainder \text{ } of \text{ } 1$$
> $$1 \div 2 = 0 \text{ with a } remainder \text{ } of \text{ } 1$$
>
> The answer is binary 1100. This equals $8 + 4 + 0 + 0$ for decimal 12.

Practice Problems 12-F
Answers at End of Chapter

Convert to binary form.

1.	2	**5.**	7	**9.**	23
2.	4	**6.**	9	**10.**	32
3.	5	**7.**	11	**11.**	63
4.	6	**8.**	16	**12.**	64

12-6 Binary-Coded Decimal (BCD) Numbers

A number code is a system of rules for converting from one form to another. The purpose is usually to make calculations more convenient. In the BCD system, each digit from 0 to 9 in a decimal number is coded as an equivalent four place binary digit number. A *bi*nary digi*t* is called a *bit*. A number made up of four bits is called a *nibble*. The nibbles are in the same order as the corresponding decimal places. All the possibilities for the decimal digits and binary nibbles are

Decimal Digit	Nibble	Decimal Digit	Nibble
0	0000	5	0101
1	0001	6	0110
2	0010	7	0111
3	0011	8	1000
4	0100	9	1001

As an example, the decimal number 39 can be converted to 0011 1001 in BCD form. The 0011 corresponds to 3, and the 1001 is 9. For a conversion the opposite way, the BCD value of 1001 0010 corresponds to 92. More nibbles can be used for additional decimal places.

The BCD conversion for 92 can be illustrated as follows:

$$\overline{\underset{1001}{}\overset{9}{}}\quad\overline{\underset{0010}{}\overset{2}{}}$$

Each binary nibble is a separate value with four bits. Do not count more than four binary places in the BCD system, as the highest possible value for each nibble is 9.

Such BCD conversions are useful for the input and output of an electronic calculator. You punch in the decimal numbers. These are converted to binary form for the digital circuits. After the calculations are processed in binary form, they are converted to decimal numbers for the display. It should be noted that the BCD system is only one of many types of number codes that can be used to simplify operations in many digital devices.

Practice Problems 12-G
Answers at End of Chapter

Convert these decimal numbers to BCD form.

1.	3	**5.**	33
2.	8	**6.**	69
3.	16	**7.**	85
4.	22	**8.**	97

Convert these BCD numbers to decimal numbers.

9.	0011	**13.**	0011 0011
10.	1000	**14.**	0110 1001
11.	0001 0110	**15.**	1000 0101
12.	0010 0010	**16.**	1001 0111

12-7 Octal Number System

Octal means eight. The base or radix of the octal number system is 8. Place values for octal numbers are in powers of 8. Illustrated for just three places, the values are

$$8^2 \qquad 8^1 \qquad 8^0 \text{ (Units)}$$

$$64s \qquad 8s \qquad 0 \text{ to } 7$$

Multiply each place value to the left by 8 for the next higher place value. In any place, the digit for the count can be 0 to 7. As an example, $(127)_8$ is equal to decimal 87. The octal place counts are

$$(1 \times 64) + (2 \times 8) + 7$$
$$64 \quad + \quad 16 \quad + 7 = 87$$

The advantage of octal numbers over binary numbers is that they need fewer places than binary numbers for the same value. For instance, $(87)_{10}$ is 0101 0111 in binary and 127 in octal.

Practice Problems 12-H
Answers at End of Chapter

Convert the following octal numbers to decimal numbers.

1.	$(005)_8$	**5.**	$(123)_8$
2.	$(011)_8$	**6.**	$(132)_8$
3.	$(111)_8$	**7.**	$(147)_8$
4.	$(226)_8$	**8.**	$(777)_8$

12-8 Hexadecimal System

Hexadecimal means six (hexa) and ten (deci) combined, or sixteen. The base or radix is 16 in the hexadecimal system. This notation is usually referred to as *hex*. In order to have 16 characters for each place or position in a hex number, the ten digits 0 to 9 are used along with the six capital letters A, B, C, D, E, and F for decimal numbers 10 to 15. The hex values with binary and decimal equivalents are listed in Table 12-1 on the next page. The hex system does not use the letter O, so that hex 0 means the digit zero. The 0 is often written with a slash: Ø, to avoid confusion with O.

Table 12-1 Hexadecimal Equivalents

(Hexadecimal)$_{16}$	(Binary)$_2$	(Decimal)$_{10}$
0	0000	0
1	0001	1
2	0010	2
3	0011	3
4	0100	4
5	0101	5
6	0110	6
7	0111	7
8	1000	8
9	1001	9
A	1010	10
B	1011	11
C	1100	12
D	1101	13
E	1110	14
F	1111	15

The hexadecimal system is generally used in computers because hex values are much shorter than binary numbers. The reason is that a base of 16 is larger than 2, so that the same numerical value requires fewer places. Furthermore, it is relatively easy to convert between hex and binary notation. The digital circuits in a computer work only with binary values, but inputs and outputs are easier to visualize without long strings of 1s and 0s.

Practice Problems 12-I
Answers at End of Chapter

Convert the following decimal values to hex notation:

1.	5	**5.**	12
2.	9	**6.**	13
3.	10	**7.**	14
4.	11	**8.**	15

Consider a hex value with two characters, such as 3F. To convert to binary, just change each character to a binary nibble, with corresponding positions for the place values. Then

$$3F = 0011\ 1111$$

The 3 is binary 0011 or $2 + 1 = 3$. The hex F for decimal 15 is binary 1111 or $8 + 4 + 2 + 1 = 15$. Note

that the two-position hex value is equivalent to an 8-bit binary number. An 8-bit binary number is called a *byte*.

The decimal count for hex 3F can be calculated either from the binary form or directly from the hex values. In binary form, the hex 3F is 0011 1111. The decimal count for the binary 0011 1111 is

$$0 + 0 + 32 + 16 + 8 + 4 + 2 + 1 = 63$$

Consider the decimal values for each place or position in a hex value. These decimal counts, illustrated for just four positions, are

$$16^3 \qquad 16^2 \qquad 16^1 \qquad 16^0 \text{ (units)}$$
$$4096s \qquad 256s \qquad 16s \qquad 0 \text{ to } 15$$

The place values to the left increase in multiples of 16. For the hex value 3F, as an example, the decimal count is

$$(3 \times 16) + 15$$
$$\text{or} \qquad 48 \quad + 15 = 63$$

This count of 63 for hex 3F is the same 63 as binary 0011 1111.

Hex values are usually written in pairs, such as 3F, 07, 23, or AA for two places and 376D or 51A6 for four places. This way each pair of hex characters corresponds to a binary byte with 8 bits. As a review of the hex values and to illustrate examples of the corresponding binary bytes, note the following conversions for hex values with two characters.

$$(9)_{10} = (09)_{16} = 0000\ 1001$$
$$(10)_{10} = (0A)_{16} = 0000\ 1010$$
$$(11)_{10} = (0B)_{16} = 0000\ 1011$$
$$(12)_{10} = (0C)_{16} = 0000\ 1100$$
$$(13)_{10} = (0D)_{16} = 0000\ 1101$$
$$(14)_{10} = (0E)_{16} = 0000\ 1110$$
$$(15)_{10} = (0F)_{16} = 0000\ 1111$$
$$(16)_{10} = (10)_{16} = 0001\ 0000$$
$$(17)_{10} = (11)_{16} = 0001\ 0001$$
$$(31)_{10} = (1F)_{16} = 0001\ 1111$$
$$(32)_{10} = (20)_{16} = 0010\ 0000$$
$$(33)_{10} = (21)_{16} = 0010\ 0001$$
$$(42)_{10} = (2A)_{16} = 0010\ 1010$$
$$(64)_{10} = (40)_{16} = 0100\ 0000$$

Decimal numbers greater than 15 can be converted to hex values by the division method used for binary conversion (the double-dabble method). Instead of dividing by 2, though, divide by 16 for hex conversion. Divide successively until the quotient is zero and use the remainders for the hex characters. As an example, to convert decimal 18 to hex values,

$$18 \div 16 = 1 \text{ with a } remainder \text{ } of \text{ } 2$$
$$1 \div 16 = 0 \text{ with a } remainder \text{ } of \text{ } 1$$
$$\text{hex number } 1 \quad 2$$

The hex value then is $(12)_{16}$.

When the remainder is more than 9, use the letters A to F for the equivalent hex character.

Example   Convert decimal 63 to hex values.

Answer

$$63 \div 16 = 3 \text{ with a } remainder \text{ } of \text{ } 15 \text{ } or \text{ } F$$
$$3 \div 16 = 0 \text{ with a } remainder \text{ } of \text{ } 3$$

The hex value for 63 then is 3F. As a check on the decimal count, this value is

$$(3 \times 16) + 15$$
$$48 \quad + 15 = 63$$

Practice Problems 12-J
Answers at End of Chapter

Convert the following decimal numbers to hex form with two characters and equivalent binary values as 8-bit bytes.

1.	7	**6.**	33
2.	11	**7.**	64
3.	15	**8.**	67
4.	16	**9.**	240
5.	17	**10.**	255

Give the decimal numbers for the following hex values.

11.	07	**16.**	21
12.	0B	**17.**	40
13.	0F	**18.**	43
14.	10	**19.**	F0
15.	11	**20.**	FF

When hex values have four characters, the equivalent binary number has 16 bits in 2 bytes. As an example, consider $(123F)_{16}$. The binary equivalent is

$$\underbrace{0001 \ 0010}_{(12)_{16}} \qquad \underbrace{0011 \ 1111}_{(3F)_{16}}$$

For these long binary numbers, it is convenient to consider the 16 bits in two parts, called the upper byte (UB) and lower byte (LB). The upper byte has the higher place values. For this example, the lower byte is

$$LB = 0011 \ 1111$$

The decimal count for these binary places, from left to right, equals $0 + 0 + 32 + 16 + 8 + 4 + 2 + 1 = 63$. In addition, the upper byte here is

$$UB = 0001 \ 0010$$

It is important to note that the place values for the upper byte start from 256 or 2^8 for the place at the right and increase in multiples of 2 to 32,768 or 2^{15} for the place at the left. The decimal count for these places from left to right in the UB of 0001 0010 is $0 + 0 + 0 + 4096 + 0 + 0 + 512 + 0 = 4608$. The 512 is 2^9 and the 4096 is 2^{12}.

The total decimal value is the sum of the lower and upper bytes. For this example,

$$LB + UB = 63 + 4608$$
$$= 4671$$

This value of 4671 is the decimal value of hex 123F.

The hex value can also be converted to a decimal count directly. In powers of 16, the value of $(123F)_{16}$ becomes

$$(1 \times 4096) + (2 \times 256) + (3 \times 16) + 15$$
$$4096 \quad + \quad 512 \quad + \quad 48 \quad + 15 = 4671$$

Note that this sum of 4671 is the same decimal value obtained by adding the upper and lower bytes in binary form.

Practice Problems 12-K
Answers at End of Chapter

Convert the following hex values to binary form.

1.	03	7.	F0
2.	1F	8.	9A
3.	20	9.	EE
4.	2A	10.	031F
5.	78	11.	202A
6.	40	12.	1A1B

Practice Problems 12-L
Answers at End of Chapter

Convert the following binary values to hex values.

1.	0000 0011	7.	1111 0000
2.	0001 1111	8.	1001 1010
3.	0010 0000	9.	1110 1110
4.	0010 1010	10.	0000 0011 0001 1111
5.	0111 1000	11.	0010 0000 0010 1010
6.	0100 0000	12.	0001 1010 0001 1011

Practice Problems 12-M
Answers at End of Chapter

Convert the following hex values to decimal numbers.

1.	03	7.	F0
2.	1F	8.	9A
3.	20	9.	EE
4.	2A	10.	031F
5.	78	11.	202A
6.	40	12.	1A1B

Convert the following binary values to decimal numbers.

13.	0000 0011	19.	1111 0000
14.	0001 1111	20.	1001 1010
15.	0010 0000	21.	1110 1110
16.	0010 1010	22.	0000 0011 0001 1111
17.	0111 1000	23.	0010 0000 0010 1010
18.	0100 0000	24.	0001 1010 0001 1011

12-9 The ASCII Code

The abbreviation ASCII (pronounced *ask-key*) stands for American Standard Code for Information Interchange. The code is used to convert digits, letters, symbols, and punctuation marks into binary form for the input to computers. You punch in the information on the keyboard and each key produces the corresponding binary signal. The combination of numbers, letters, and symbols is called *alphanumeric* information.

The ASCII code uses 7 bits to identify each alphanumeric character, as shown in Table 12-2. Actually, the binary information is processed as 8-bit bytes, with one bit reserved for checking errors.

Table 12-2 ASCII Code

$X_3X_2X_1X_0$	$X_6X_5X_4$					
	010	**011**	**100**	**101**	**110**	**111**
0000	SP	0	@	P		p
0001	!	1	A	Q	a	q
0010	"	2	B	R	b	r
0011	#	3	C	S	c	s
0100	$	4	D	T	d	t
0101	%	5	E	U	e	u
0110	&	6	F	V	f	v
0111	'	7	G	W	g	w
1000	(	8	H	X	h	x
1001	)	9	I	Y	i	y
1010	*	:	J	Z	j	z
1011	+	;	K		k	
1100	,	<	L		l	
1101	–	=	M		m	
1110	.	>	N		n	
1111	/	?	O		o	

Table 12-2 lists the 7-bit coded values in groups of 4 bits and 3 bits. Consider the 7 bits in the following sequence:

$$X_6\ X_5\ X_4 \quad X_3\ X_2\ X_1\ X_0$$

where each X represents a binary 1 or 0. Read the bits left to right from X_6 to X_0. As an example, capital letter P corresponds to

101 0000

The 101 comes from the fourth column of $X_6\ X_5\ X_4$ values, and the 0000 is at the top of the column for $X_3\ X_2\ X_1\ X_0$ values.

In the column of 010 values in the $X_6\ X_5\ X_4$ places, all the characters are punctuation marks and symbols.

The SP stands for a space in typing. The ASCII code for the space is 010 0000.

In the next column, 011 values for the $X_6 X_5 X_4$ places, the characters include the digits 0 to 9. For example, 9 is coded as 011 1001. Note that 1001 is the BCD value for 9.

The remaining columns show that all capital letters have 100 or 101 for the $X_6 X_5 X_4$ places. Also all lower-case (small) letters use 110 or 111. In general, the binary digits in the three places for $X_6 X_5 X_4$ indicate what type of character will be in the places for $X_3 X_2 X_1 X_0$.

Practice Problems 12-N
Answers at End of Chapter

Convert the following characters to the ASCII code.

1.	B	**5.**	3
2.	b	**6.**	8
3.	G	**7.**	*
4.	m	**8.**	=

Convert the following binary values in ASCII code to the equivalent characters.

9.	100 0010	**13.**	011 0011
10.	110 0010	**14.**	011 1000
11.	100 0111	**15.**	010 1010
12.	110 1101	**16.**	011 1101

12-10 Binary Logic

Many mechanical and electrical systems have only two states of operation. For example, a light switch can be ON or OFF, a valve can be OPEN or CLOSED, and a drawbridge can be UP or DOWN. In each case the two states could be compared to a binary system. The ON position of the switch, the OPEN position of the valve, and the DOWN position of the drawbridge can be given the value 1. The OFF position of the switch, the CLOSED position of the valve, and the UP position of the drawbridge could be considered to have the value 0.

Imagine a light bulb connected to a battery through a switch as shown in Fig. 12-2. When the switch has a value 1 (ON) the light bulb is lit. When the switch has a value of 0 (OFF) the bulb is unlit. If a second switch was now added to this system the effect of each individual switch would be changed. In Fig. 12-3a, on the next page, the two switches are connected in tandem, that is, one directly in line with the other. This is usually

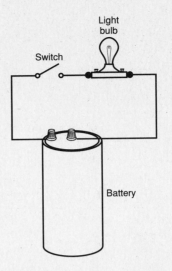

Fig. 12-2 A switching circuit for a light bulb.

called a *series* connection. In order for the bulb to light, a direct unbroken path must exist from the battery through the switches and the bulb and back to the battery. If either switch has the value 0 (OFF), the light bulb will be unlit. Only when both switches have the value 1 (both ON) will the bulb light. In Fig. 12-3b the switches are connected across one another and then connected to the light bulb. This is usually called a *parallel* connection. In this case the switches present two separate independent paths from the battery through the switch and bulb and back to the battery. In other words, there are two parallel paths in this arrangement. In this case the path can be completed if either one of the switches has the value 1 (ON).

Because of the way they can control the operation or condition of a system, the combination of switches in Fig. 12-3 are called *logic circuits* or *logic gates*.

Practice Problems 12-O
Answers at End of Chapter

For each of the following cases give the state of the device described (1 or 0).

1. The bulb is lit. The light switch value is __1__ .
2. Water is flowing in the pipe. The water valve value is __1__ .
3. Traffic is stopped while the drawbridge is up. The drawbridge value is __0__ .

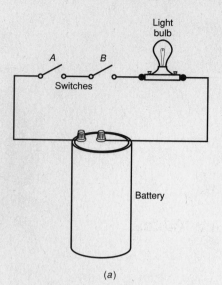

Light bulb

A   B

Switches

Battery

(a)

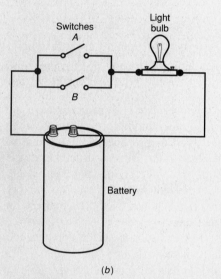

Switches
A

Light bulb

B

Battery

(b)

Fig. 12-3 Two switches control a light bulb. (a) Switches connected in series. (b) Switches connected in parallel.

12-11 AND Gate

Consider two cities separated by a wide river in the middle of which is an island. Each city is connected to the island by a drawbridge. In order to get from one city to another both bridges must be down (Fig. 12-4). We can denote the bridges as A and B and the condition for crossing from one city to another, Y. We can then say that the position of A (UP or DOWN) and the position of B (UP or DOWN) equals the ability to get from one city to another (Y). The arrangement of bridges is an example of AND logic; the two bridges comprise a logic circuit known as an AND gate. Travel from one city to another is possible only if bridge A *and* bridge B are down.

A similar AND gate using switches was shown in Fig. 12-3a. The mathematical expression describing this arrangement is

$$A \cdot B = Y$$

This says that the output Y (the light ON or OFF) depends on the condition of switch A AND switch B. The dot between A and B denotes the logic operation AND. If A is open (0) and B is closed (1) the path will not be complete and there will not be an output at Y (denoted by 0).

$$0 \cdot 1 = 0$$

If both switches are closed (both 1s) there will be a complete path and the output would be 1.

$$1 \cdot 1 = 1$$

Remember that in logic circuits the dot · denotes AND; it does not indicate multiplication.

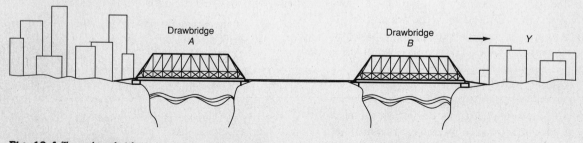

Drawbridge
A

Drawbridge
B

Y

Fig. 12-4 Two drawbridges in series control traffic.

Although the examples given above use only two switches, AND circuits may have many more. In the expression $A \cdot B \cdot C$ there are three switches in series. All three must be 1 in order for the output to be 1. If A, B, or C have the value 0, the output will be 0.

Practice Problems 12-P
Answers at End of Chapter

State whether the output is 1 or 0 for each of the following.

1.	$1 \cdot 1 =$ 1	**6.**	$A \cdot 0 =$ 0
2.	$0 \cdot 0 =$ 0	**7.**	$B \cdot 0 =$ 0
3.	$0 \cdot 0 \cdot 1 =$ 0	**8.**	$A \cdot B \cdot 0 =$ 0
4.	$1 \cdot 1 \cdot 1 \cdot 1 =$ 1	**9.**	$a \cdot 0 \cdot B =$ 0
5.	$1 \cdot 0 =$ 0	**10.**	$1 \cdot A \cdot 0 =$ 0

12-12   OR Gate

The two switches A and B in Sec. 12-11 can be arranged in parallel and connected to one output, Y, as in Fig. 12-3b. In this arrangement, each switch can complete the path to the output Y. This can be expressed mathematically as

$$A + B = Y$$

This says that the output Y depends on the state of A OR B. The plus sign in this case does not indicate addition; it is the symbol for the OR logic function. If A is ON (1) and B is OFF (0), the output Y would be ON (1),

$$1 + 0 = 1$$

which is read: 1 OR 0 equals 1. If A is OFF (0) and B is ON (1), the output would, again, be ON (1).

$$0 + 1 = 1$$

Although the logic expression looks like a similar arithmetic expression, in the case above the plus sign is a symbol for OR logic, not addition.

The examples given above use only two switches, but an OR gate can have any number of switches (or other 2-state devices) in parallel. The same condition of the OR gate holds true: if any one of the switches has a value 1, the output will be 1.

Practice Problems 12-Q
Answers at End of Chapter

State whether the output is 1 or 0 for each of the following.

1.	$1 + 1 =$ 1	**6.**	$A + 1 =$ 1
2.	$0 + 1 =$ 1	**7.**	$B + 1 =$ 1
3.	$0 + 0 + 1 =$ 1	**8.**	$A + B + 1 =$ 1
4.	$1 + 1 + 1 + 1 =$ 1	**9.**	$A + 1 + B =$ 1
5.	$1 + 0 =$ 1	**10.**	$1 + A + 0 =$ 1

12-13   Truth Tables

In the discussion of the OR and AND logic gates in Secs. 12-11 and 12-12 reference was made to the possible combinations of ON (or closed) and OFF (or open) switches. All of the possible combinations and their outputs can be tabulated in what are called *truth tables*. Figure 12-5 shows the truth tables for the AND logic gate using two switches. Four combinations are possible for the two switches. Figure 12-6a, on the next page, shows the state of the switches and the resulting output while Fig. 12-6b shows the binary values for each case. When both A and B are 0, the output Y will be 0. When A is 1 and B is 0, Y is 0. The output Y can be 1 only when both A AND B are 1.

$$A \cdot B = Y$$

A	B	Y
0	0	0
0	1	0
1	0	0
1	1	1

Fig. 12-5 Truth table for a two-input AND gate.

Figure 12-7 shows the truth tables for the OR logic gate using two switches. If either A OR B is 1 (ON), the output will be 1 (ON). The only combination producing no output (that is, $Y = 0$) is both A and B equal to 0.

Truth tables can be constructed for more than two switches but the number of combinations increases accordingly. With two switches there are 2^2 or 4 combinations; with three switches there are 2^3 or 8 combinations; with four switches there are 2^4 or sixteen combinations. The number of combinations is equal to 2^n where n is the number of switches in the logic gate.

Practice Problems 12-R
Answers at End of Chapter

1. Complete the truth table of Fig. 12-8.
2. Complete the truth table of Fig. 12-9.

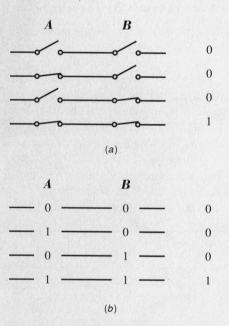

A   *B*

(a)

A   *B*

(b)

Fig. 12-6 Switch combinations in series.
(a) ON-OFF positions. (b) Binary values.

$$A + B = Y$$

A	*B*	*Y*
0	0	0
0	1	1
1	0	1
1	1	1

Fig. 12-7 Truth table for a two-input OR gate.

$$A + B + C = Y$$

A	*B*	*C*	*Y*
0	0	0	0
0	0	1	1
0	1	0	1
0	1	1	1
1	0	0	1
1	0	1	1
1	1	0	1
1	1	1	1

Fig. 12-8 Truth table for Practice Problem 12-R-1.

$$A \cdot B \cdot C = Y$$

A	*B*	*C*	*Y*
0	0	0	0
0	0	1	0
0	1	0	0
0	1	1	0
1	0	0	0
1	0	1	0
1	1	0	0
1	1	1	1

Fig. 12-9 Truth table for Practice Problem 12-R-2.

12-14 Logic Symbols

Up to this point we have discussed logic circuits, or gates, in terms of switches. The two states of the switches were ON (1) or OFF (0). In digital logic, use is made of certain standard symbols representing the switching, or logic, function. Figure 12-10 shows the symbol for an AND gate. The two lines marked *A* and *B* represent the inputs and may be compared to the condition of switches that have the values of 1 or 0. The other line marked *Y* is the output. This may be compared to the state of the path resulting from the conditions of *A* and *B*. If *A* and *B* are both 1, then *Y* will be 1. If *A* or *B* is not 1, then *Y* will be 0. The truth table for

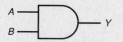

Fig. 12-10 AND gate logic symbol.

the AND gate demonstrating this is covered in Sec. 12-13.

Figure 12-11 is the standard logic symbol for an OR gate. Again A and B represent inputs and Y represents the output resulting from the states of A and B. Later sections will discuss other gates and their logic symbols.

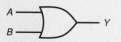

Fig. 12-11 OR gate logic symbol.

Although only two input lines are shown there may be many more than two. Each line would be identified with its own letter: A, B, C, and so forth. In all cases there is only one output line.

Symbols of different gates may be combined with one another to represent complex logic expressions. The output of one gate could be the input of another gate. The final output is a result of all gates acting together.

Practice Problems 12-S
Answers at End of Chapter

1. Construct the truth table for the logic circuit shown in Fig. 12-12.

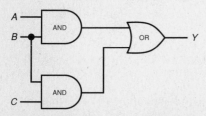

Fig. 12-12 Logic circuit for Practice Problem 12-S-1.

2. With the gates reversed so that each AND gate is an OR gate and the OR gate is an AND gate, construct the truth table.

12-15 The NOT or Inverter

The NOT, or inverter, gate changes the condition at its input to the complement at its output. For example, a NOT gate will change a 1 at its input to a 0 at its output or a 0 to a 1.

The symbol for the NOT gate is shown in Fig. 12-13. If the input to the NOT gate is A, the output will be the complement of A, written $\overline{A}$ (read "A bar"). If the input to the NOT gate is 1 the output will be 0 and vice versa. Notice that the NOT gate contains only one input and one output. The circle at the tip of the triangle is a standard symbol denoting NOT. It is used in conjunction with OR and AND gates as will be seen in the next section.

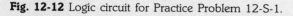

Fig. 12-13 NOT gate logic symbol.

Practice Problems 12-T
Answers at End of Chapter

1. If the input to an inverter is B, what is the output? $\overline{B}$
2. If the input to an inverter is $\overline{D}$, what is the output? D
3. Write the expression that describes the action of the NOT gate. $A = \overline{A}$ ꟾNVERTER

12-16 Other Logic Gates

Although the AND, OR, and NOT gates are the basic building blocks of logic circuits, other gates are also extremely important.

Other gates combine the actions of the three basic gates. The AND gate can be combined with the NOT gate so that the output of the AND gate is inverted for all combinations of inputs. The gate that performs this function is the NAND gate (from NOT AND). Figure 12-14, on the next page, shows the logic symbol and truth table for the NAND gate. Notice the circle at the output indicating NOT, or inversion.

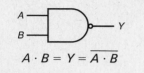

$$A \cdot B = Y = \overline{A \cdot B}$$

A	B	Y
0	0	1
0	1	1
1	0	1
1	1	0

Fig. 12-14 NAND gate logic symbol and truth table.

The OR gate can be combined with the NOT gate to produce a NOR gate (from NOT OR). Figure 12-15 shows the logic symbol and truth table for the NOR gate.

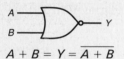

$$A + B = Y = \overline{A + B}$$

A	B	Y
0	0	1
0	1	0
1	0	0
1	1	0

Fig. 12-15 NOR gate logic symbol and truth table.

Recall that the OR gate had an output value 1 when any one or more, including all, inputs were 1. There is a logic gate that produces a 1 only if an odd number of inputs are 1 but produces a 0 value when an even number of inputs are 1. In a two-input gate this means either A or B must be 1 to produce an output 1. If both A and B are 1 the output will be 0. This logic gate is called the *Exclusive* OR gate, shortened to XOR gate. To distinguish this operation from the usual OR function, +, the symbol ⊕ is used. Figure 12-16 shows the symbol and truth table for the XOR gate.

The Exclusive NOR gate, called the XNOR gate, is shown in Fig. 12-17 together with its truth table.

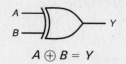

$$A \oplus B = Y$$

A	B	Y
0	0	0
0	1	1
1	0	1
1	1	0

Fig. 12-16 XOR gate logic symbol and truth table.

While these seven gates provide the full range of logic functions, their individual functions can be constructed from a combination of the others. The NAND gate, in particular, can be connected to other NAND gates to produce the same results as the AND, OR, NOT, NOR, XOR, and XNOR gates.

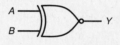

$$A \oplus B = Y = \overline{A \oplus B}$$

A	B	Y
0	0	1
0	1	0
1	0	0
1	1	1

Fig. 12-17 XNOR gate logic symbol and truth table.

Practice Problems 12-U
Answers at End of Chapter

1. Identify the gate whose function is being duplicated by the arrangement of NAND gates in Fig. 12-18a.
2. Identify the gate whose function is being duplicated by the arrangement of NAND gates in Fig. 12-18b.
3. If A is 1 in Fig. 12-18c, what is the output Y?

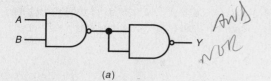

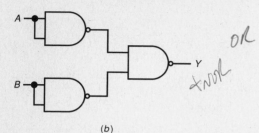

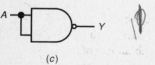

Fig. 12-18 Logic circuits for Practice Problems 12-U-1 to 12-U-3.

12-17 Logic Theorems

By applying some basic operations of algebra to boolean expressions the expressions can be simplified. The simplification will also reduce the number of logic gates required to perform a particular operation. Simplification can be done by using logic theorems. Theorems are statements that can be proved from established facts or formulas. In this section we will use thirteen basic theorems; a number of them will also be proven here.

The letters A, B, C, and D are used, as before, to denote inputs to the logic gates or switches. The binary numbers 1 and 0 have their usual meaning.

The theorems are as follows:

1. $A + 1 = 1$
2. $A + 0 = A$
3. $A \cdot 1 = A$
4. $A \cdot 0 = 0$
5. $A + A = A$
6. $A \cdot \overline{A} = 0$
7. $A + \overline{A} = 1$
8. $A \cdot (B + C) = A \cdot B + A \cdot C$
9. $(A + B) \cdot (C + D) = A \cdot C + B \cdot C + A \cdot D + B \cdot D$
10. $A \cdot B + A = A$
11. $A \cdot B + (A + B) = A + B$
12. $A \cdot B + \overline{B} = A + \overline{B}$
13. $A \cdot \overline{B} + B = A + B$

Of course, the theorems hold true for any other letters substituted for those given in the statements above. For example, theorem 5 holds true for $B + B = B$, $C + C = C$, an so on; theorem 8 can also be stated as $D \cdot (A + B) = D \cdot A + D \cdot B$; theorem 6 is also $B \cdot \overline{B} = 0$, $C \cdot \overline{C} = 0$, an so on.

The proof of some of these theorems is almost intuitive. For example, theorem 1 says that a switch in parallel with an ON (1) switch will cause the circuit to be ON (1) no matter what the condition of the first switch. Similarly theorem 4 says that a switch A in series with an OFF (0) switch will cause the circuit to be OFF (0) no matter what the condition of A.

One method of proof is to construct the truth tables for the boolean expressions on both sides of the equal sign and verify that both sides are exactly equal. Theorem 12 is verified in Fig. 12-19 using truth tables. Notice here that the equation uses both B and the complement, or inverse, of B which is $\overline{B}$.

$$A \cdot B + \overline{B}$$

A	B	$A \cdot B$	$\overline{B}$	Y
0	0	0	1	1
0	1	0	0	0
1	0	0	1	1
1	1	1	0	1

$$A + \overline{B}$$

A	B	$\overline{B}$	Y
0	0	1	1
0	1	0	0
1	0	1	1
1	1	0	1

Fig. 12-19 Verification of theorem 12 using truth tables.

Practice Problems 12-V
Answers at End of Chapter

1. Prove theorem 8.
2. Find $\overline{B} \cdot \overline{B}$.
3. Prove theorem 11.

12-18 De Morgan's Theorem

De Morgan's theorem states that:

If an OR expression is replaced by an AND expression whose elements are complements, or inverts, of the elements in the OR expression, the result will be a complement of the original expression.

An example will show how De Morgan's theorem is applied. In the expression $\overline{A} \cdot \overline{B}$ we can convert $\overline{A}$ to A and $\overline{B}$ to B. The original expression was an OR, denoted by the dot. If we now make it an AND expression, the new expression will be

$$A + B$$

By De Morgan's theorem, this is the complement of $\overline{A} \cdot \overline{B}$ or

$$\overline{A + B} = \overline{A} \cdot \overline{B}$$

The truth table of $\overline{A + B}$ and $\overline{A} \cdot \overline{B}$ (Fig. 12-20) will verify that the two expressions are equal.

In some cases a series of OR expressions may be easier to work with than AND expressions. De Morgan's theorem can help convert from one to the other.

For example $A \cdot B \cdot C$ can be changed to OR by finding the complement of each term and changing AND to OR.

$$\overline{A} + \overline{B} + \overline{C}$$

This new expression is the complement of the original expression, that is,

$$\overline{A} + \overline{B} + \overline{C} = \overline{A \cdot B \cdot C}$$

Practice Problems 12-W
Answers at End of Chapter

Apply De Morgan's theorem to find the complement of each of the following expressions.

1. $A \cdot \overline{B}$ $\overline{A} + B$
2. $\overline{B + A}$ $\overline{B} + \overline{A}$
3. $A + B \cdot \overline{C}$ $\overline{A} + \overline{B} + C$
4. $A \cdot \overline{B} + C$ $\overline{A} + B + \overline{C}$
5. $\overline{A} + \overline{B} + C$ $A + B + \overline{C}$

12-19 Simplifying Boolean Expressions

Complex boolean expressions can be reduced, sometimes considerably, by applying the theorems given in Secs. 12-17 and 12-18.

The expression

$$A \cdot (C + \overline{C}) + \overline{B} \cdot (A + \overline{B}) + C \cdot (D + \overline{D}) = Y$$

contains 9 separate elements, and six individual inputs, A, $\overline{B}$, C, $\overline{C}$, D, and $\overline{D}$. The problem is to eliminate as

$\overline{A + B}$

A	B	$A + B$	Y
0	0	0	1
0	1	1	0
1	0	1	0
1	1	1	0

$\overline{A} \cdot \overline{B}$

A	B	$\overline{A}$	$\overline{B}$	Y
0	0	1	1	1
0	1	1	0	0
1	0	0	1	0
1	1	0	0	0

Fig. 12-20 De Morgan's theorem verified using truth tables.

many inputs (or switches) as possible and still have the expression produce the same output Y. Remember the bar over a letter indicates a NOT gate which results in a reversal of the given condition. If $B = 0$, then $\overline{B} = 1$ (and vice versa).

Theorem 7 can simplify the first and last terms since

$$C + \overline{C} = 1 \text{ and } D + \overline{D} = 1$$

Thus the new expression is $A \cdot 1 + \overline{B} \cdot A + \overline{B} + C \cdot 1$ (since $\overline{B} \cdot \overline{B} = \overline{B}$, $A \cdot A = A$, and so on.)

By theorems 3 and 5

$$A + \overline{B} \cdot A + \overline{B} + C = Y$$

combining the A terms

$$A \cdot (1 + \overline{B}) + \overline{B} + C = Y$$

But theorem 1 says $1 + \overline{B} = 1$. Therefore

$$A + \overline{B} + C = Y$$

This final expression, using only three switches, or inputs, performs the same functions as the original expression with nine elements. The truth tables in Fig. 12-21 verifies that these two expressions are equal.

$$A \cdot (C + \overline{C}) + \overline{B} \cdot (A + \overline{B}) + C \cdot (D + \overline{D})$$

A	B	C	$C + \overline{C}$	$\overline{B}$	$A + \overline{B}$	$D + \overline{D}$	$A \cdot (C + \overline{C})$	$\overline{B} \cdot (A + \overline{B})$	$C \cdot (D + \overline{D})$	Y
0	0	0	1	1	1	1	0	1	0	1
0	0	1	1	1	1	1	0	1	1	1
0	1	0	1	0	0	1	0	0	0	0
0	1	1	1	0	0	1	0	0	1	1
1	0	0	1	1	1	1	1	1	0	1
1	0	1	1	1	1	1	1	1	1	1
1	1	0	1	0	1	1	1	0	0	1
1	1	1	1	0	1	1	1	0	1	1

$$A + \overline{B} + C$$

A	B	C	$\overline{B}$	Y
0	0	0	1	1
0	0	1	1	1
0	1	0	0	0
0	1	1	0	1
1	0	0	1	1
1	0	1	1	1
1	1	0	0	1
1	1	1	0	1

Fig. 12-21 Verification of $A \cdot (C + \overline{C}) + \overline{B} \cdot (A + \overline{B}) + C \cdot (D + \overline{D}) = A + \overline{B} + C$

As another example, the expression

$$A \cdot \overline{B} + \overline{A} \cdot B + \overline{A} \cdot \overline{B} = Y$$

will be simplified. Combining terms

$$A \cdot \overline{B} + \overline{A} \cdot (B + \overline{B})$$

From theorem 7, $B + \overline{B} = 1$

$$A \cdot \overline{B} + \overline{A} \cdot 1$$
$$A \cdot \overline{B} + \overline{A} = \overline{B} \cdot A + \overline{A}$$

Applying theorem 12,

$$\overline{B} \cdot A + \overline{A} = \overline{B} + \overline{A} = \overline{A} + \overline{B}$$
or $$\overline{A} + \overline{B} = Y$$

Practice Problems 12-X
Answers at End of Chapter

1. Simplify the following boolean equation using the theorems in Section 12-18.

$$\overline{A} \cdot B + A \cdot B + A \cdot C + \overline{B} \cdot C + A \cdot B \cdot C = Y$$

2. Simplify the following equation using the theorems in Section 12-18. Verify your answer using truth tables.

$$(A + B + C) \cdot (A + B + C) = Y$$

Review Problems
Answers to Odd-Numbered Problems at Back of Book

The following problems review the methods of computer mathematics covered in this chapter.

1. $(0011)_2 = (\ 03\ )_{10}$ ✓
2. $(0011)_8 = (\qquad)_{10}$
3. $(2F)_{16} = (\qquad)_2 = (\qquad)_{10}$ ✓
4. $(023)_8 = (\qquad)_{10} = (\qquad)_{16}$
5. $(1111)_2 + (0011)_2 = (\qquad)_2$
6. $(0011)_2 - (1111)_2 = (\qquad)_2$
7. $(1111\ 1111)_2 = (\qquad)_{10} = (\qquad)_8$
8. $(FF)_{16} = (\qquad)_{10} = (\qquad)_2$
9. ASCII S $= (\qquad)_2$
10. ASCII + $= (\qquad)_2 = (\qquad)_{10}$

11. State whether the output in the following examples is 1 or 0:
    a. $0 \cdot D = 0$
    b. $(1 + A) \cdot 0 = 0$
    c. $(0 \cdot A) + 1 = 1$
12. State whether each of the following boolean equations is true or false:
    a. $A + A + 0 = 2A$
    b. $B \cdot B = B^2$
    c. $\overline{C \cdot D} = \overline{C} + \overline{D}$
13. Draw a truth table for the following boolean equation: $(A + B) \cdot (\overline{A} + \overline{B}) = Y$
14. Draw a truth table for the following boolean equation:
    $$(\overline{A \cdot B}) + (\overline{A} \cdot \overline{B}) = Y$$
15. Find the output for the logic diagram in Fig. 12-22 for each of the conditions given:
    a. $A = 1, B = 1\ Y =$ 0
    b. $A = 0, B = 0\ Y =$ 0
    c. $A = 1, B = 0\ Y =$ 0

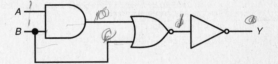

Fig. 12-22 Logic diagram for Review Problem 15.

16. Find the output for the logic diagram in Fig. 12-23 for each of the conditions given:
    a. $A = 1, B = 1\ Y =$ 0
    b. $A = 0, B = 0\ Y =$ 0
    c. $A = 0, B = 1\ Y =$ 0

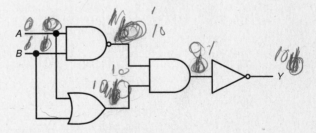

Fig. 12-23 Logic diagram for Review Problem 16.

17. Simplify the following boolean expression:
    $A \cdot (B + A) + B + A \cdot (B + A)$ 1
18. Simplify the following boolean expression:
    $B \cdot (B + A) + A + (B + A) \cdot (A \cdot B)$ 0

Answers to Practice Problems

12-A
1. Binary
2. OFF
3. Digital

12-B
1. 0
2. 1
3. 2
4. 4
5. 8
6. 12
7. 14
8. 15
9. 7
10. 3

12-C
1. $(1111)_2$ or 15
2. $(1001)_2$ or 9
3. $(0011)_2$ or 3
4. $(1111)_2$ or 15
5. $(10010)_2$ or 18
6. $(10110)_2$ or 22
7. $(10000)_2$ or 16
8. $(1110)_2$ or 14
9. $(11001)_2$ or 25
10. $(11110)_2$ or 30

12-D
1. 1111
2. 0000
3. 1010
4. 0101
5. 0110
6. 1001
7. 1100
8. 0011

12-E
1. 0011 or 3
2. 0011 or 3
3. 1000 or 8
4. 0000 or 0
5. −0010 or −2
6. 1110 or 14
7. 0100 or 4
8. −0100 or −4

12-F
1. 0010
2. 0100
3. 0101
4. 0110
5. 0111
6. 1001
7. 1011
8. 0001 0000
9. 0001 0111

10. 0010 0000
11. 0011 1111
12. 0100 0000

12-G
1. 0011
2. 1000
3. 0001 0110
4. 0010 0010
5. 0011 0011
6. 0110 1001
7. 1000 0101
8. 1001 0111
9. 3
10. 8
11. 16
12. 22
13. 33
14. 69
15. 85
16. 97

12-H
1. 5
2. 9
3. 73
4. 150
5. 83
6. 90
7. 103
8. 511

12-I
1. 5
2. 9
3. A
4. B
5. C
6. D
7. E
8. F

12-J
1. $(07)_{16}$ = 0000 0111
2. $(0B)_{16}$ = 0000 1011
3. $(0F)_{16}$ = 0000 1111
4. $(10)_{16}$ = 0001 0000
5. $(11)_{16}$ = 0001 0001
6. $(21)_{16}$ = 0010 0001
7. $(40)_{16}$ = 0100 0000
8. $(43)_{16}$ = 0100 0011
9. $(F0)_{16}$ = 1111 0000
10. $(FF)_{16}$ = 1111 1111
11. 7
12. 11
13. 15

14. 16
15. 17
16. 33
17. 64
18. 67
19. 240
20. 255

12-K
1. 0000 0011
2. 0001 1111
3. 0010 0000
4. 0010 1010
5. 0111 1000
6. 0100 0000
7. 1111 0000
8. 1001 1010
9. 1110 1110
10. 0000 0011 0001 1111
11. 0010 0000 0010 1010
12. 0001 1010 0001 1011

12-L
1. 03
2. 1F
3. 20
4. 2A
5. 78
6. 40
7. F0
8. 9A
9. EE
10. 031F
11. 202A
12. 1A1B

12-M
1. 3
2. 31
3. 32
4. 42
5. 120
6. 64
7. 240
8. 154
9. 238
10. 799
11. 8234
12. 6683
13. 3
14. 31

15. 32

16. 42

17. 120

18. 64

19. 240

20. 154

21. 238

22. 799

23. 8234

24. 6683

12-N **1.** 100 0010

2. 110 0010

3. 100 0111

4. 110 1101

5. 011 0011

6. 011 1000

7. 010 1010

8. 011 1101

9. B

10. b

11. G

12. m

13. 3

14. 8

15. *

16. =

12-O **1.** 1

2. 1

3. 0

12-P **1.** 1

2. 0

3. 0

4. 1

5. 0

6. 0

7. 0

8. 0

9. 0

10. 0

12-Q **1.** 1

2. 1

3. 1

4. 1

5. 1

6. 1

7. 1

8. 1

9. 1

10. 1

12-R **1.** $Y = 1, 1, 1, 1, 1, 1, 1$

2. $Y = 0, 0, 0, 0, 0, 0, 1$

12-S **1.**

A	B	C	$A \cdot B$	$B \cdot C$	Y
0	0	0	0	0	0
0	0	1	0	0	0
0	1	0	0	0	0
0	1	1	0	1	1
1	0	0	0	0	0
1	0	1	0	0	0
1	1	0	1	0	1
1	1	1	1	1	1

2.

A	B	C	$A + B$	$B + C$	Y
0	0	0	0	0	0
0	0	1	0	1	0
0	1	0	1	1	1
0	1	1	1	1	1
1	0	0	1	0	0
1	0	1	1	1	1
1	1	0	1	1	1
1	1	1	1	1	1

12-T **1.** $\overline{B}$

2. D

3. $A = \overline{\overline{A}}$

12-U **1.** AND

2. OR

3. NOT 1, or 0

12-V **1.**

$$A \cdot (B + C)$$

A	B	C	$B + C$	Y
0	0	0	0	0
0	0	1	1	0
0	1	0	1	0
0	1	1	1	0
1	0	0	0	0
1	0	1	1	1
1	1	0	1	1
1	1	1	1	1

$$A \cdot B + A \cdot C$$

A	B	C	$A \cdot B$	$A \cdot C$	Y
0	0	0	0	0	0
0	0	1	0	0	0
0	1	0	0	0	0
0	1	1	0	0	0
1	0	0	0	0	0
1	0	1	0	1	1
1	1	0	1	0	1
1	1	1	1	1	1

2. $\overline{B} \cdot \overline{B} = \overline{B}$

3. Theorem 10:

$A \cdot B + A = A$

Theorem 11:

$A \cdot B + (A + B) = A + B$

$\underbrace{A \cdot B + A}_{A} + B = A + B$

$A + B = A + B$

12-W **1.** $\overline{A} + B$

2. $B + A$

3. $\overline{A} \cdot \overline{B} + C$

4. $\overline{A} + B \cdot \overline{C}$

5. $A \cdot B \cdot \overline{C}$

12-X **1.** $B + A \cdot C + \overline{B} \cdot C$

2. $A + B + C$

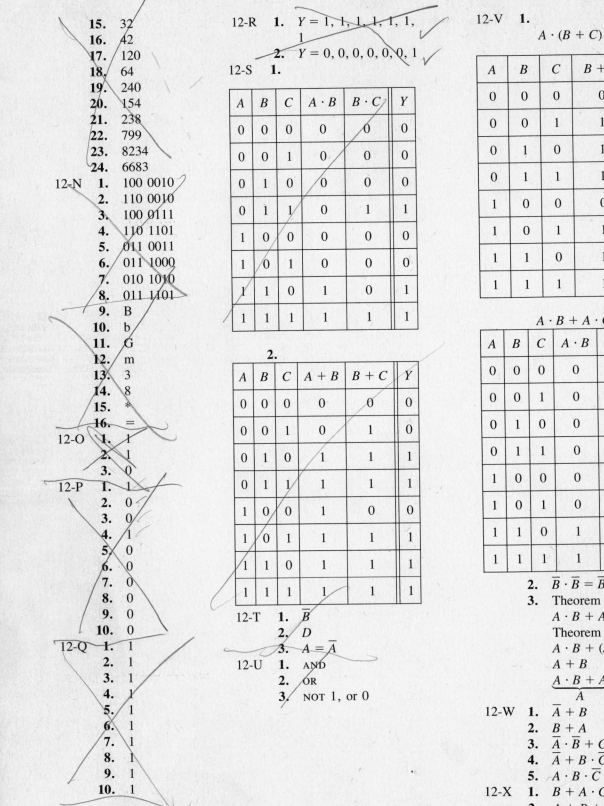

ANSWERS TO ODD-NUMBERED REVIEW PROBLEMS

Chapter 1

1.	2.847	13.	289
3.	6.195	15.	49.2
5.	0.35	17.	10.5
7.	272.8	19.	5.5
9.	7.2	21.	8%
11.	81		

Chapter 2

1.	-5	9.	-2
3.	9	13.	-0.24
5.	3.5	15.	-21
7.	24		

Chapter 3

1.	$1/16$	7.	$-1/8$
3.	4	9.	0.318
5.	$-1/3$		

Chapter 4

1.	81	11.	$1/8$
3.	13.69	13.	0.0625
5.	8	15.	9×10^8
7.	-64	17.	49
9.	625	19.	100

Chapter 5

1.	6×10^5	11.	0
3.	8×10^5	13.	9×10^6
5.	35×10^5	15.	3×10^3
	or 3.5×10^6	17.	13.2×10^6
7.	5×10^5	19.	4.2×10^3
9.	3×10^4		

Chapter 6

1.	-0.699	5.	6.806
3.	0.342	7.	4.681

9.	1.568	19.	2200
11.	2.079	21.	3 dB
13.	7.696	23.	-6 dB
15.	3.611	25.	0 dB
17.	2		

Chapter 7

1.	2.4 mA	11.	9.5 L
3.	29 kV	13.	328 cm^3
5.	22 pF	15.	44 W
7.	91.44 cm	17.	26.7°C
9.	4.545 kg	19.	419 J

Chapter 8

1.	$6a^2$	11.	$2a^2 + a$
3.	$9x^2y^2$	13.	$2y^2 + 6y$
5.	$6a^3$	15.	$9a^2 + 12ab + 4b^2$
7.	$4a/5$	17.	$9a^2 - 4b^2$
9.	a^3		

Chapter 9

1.	3	7.	8
3.	15	9.	$2a/3b$
5.	9		

Chapter 10

1.	$x = 4$	5.	$V_1 = 1$
	$y = 2$		$V_2 = -2$
3.	$I_1 = 1.5$	7.	$x = 8$
	$I_2 = -0.5$		$y = -2$

Chapter 11

1.	1.0	11.	0.332
3.	0.707	13.	1.111
5.	0.5	15.	42°
7.	0.1	17.	48°
9.	10	19.	53.1°

Chapter 12

1.	3_{10}	
3.	$(10\ 1111)_2$, 47_{10}	
5.	$(1\ 0010)_2$	
7.	255_{10}, 377_8	
9.	101 0011	
11.	a. 0	
	b. 0	
	c. 1	
13.	See table below.	
15.	a. $Y = 1$	
	b. $Y = 0$	
	c. $Y = 0$	
17.	1	

Table for Problem 13

A	B	$\overline{A}$	$\overline{B}$	$A + B$	$\overline{A} + \overline{B}$	Y
0	0	1	1	0	1	0
0	1	1	0	1	1	1
1	0	0	1	1	1	1
1	1	0	0	1	0	0

INDEX

133